CEC2C08

Communication Principles and Practices, Public Perception, and Message Effectiveness

Related to Human and Environmental Health Effects with a Focus on Trace Organic Compounds

by:

Elisabeth L. Hawley
Edward G. Means III
Rula A. Deeb
Malcolm Pirnie Inc.

Linda Macpherson
CH2M HILL

Paul Slovic
Jozee Adamson
Decision Research

2009

WERF

Water Environment Federation® Preserving & Enhancing the Global Water Environment

NACWA A Clear Commitment to America's Waters

The Water Environment Research Foundation, a not-for-profit organization, funds and manages water quality research for its subscribers through a diverse public-private partnership between municipal utilities, corporations, academia, industry, and the federal government. WERF subscribers include municipal and regional water and wastewater utilities, industrial corporations, environmental engineering firms, and others that share a commitment to cost-effective water quality solutions. WERF is dedicated to advancing science and technology addressing water quality issues as they impact water resources, the atmosphere, the lands, and quality of life.

For more information, contact:
Water Environment Research Foundation
635 Slaters Lane, Suite 300
Alexandria, VA 22314-1177
Tel: (703) 684-2470
Fax: (703) 299-0742
www.werf.org
werf@werf.org

This report was co-published by the following organization. For non-subscriber sales information, contact:

IWA Publishing
Alliance House, 12 Caxton Street
London SW1H 0QS, United Kingdom
Tel: +44 (0) 20 7654 5500
Fax: +44 (0) 20 7654 5555
www.iwapublishing.com
publications@iwap.co.uk

Library of Congress Catalog Card Number: 2008940360

IWAP ISBN: 1-84339-350-6/978-1-84339-350-4

Malcolm Pirnie Inc.

The research on which this report is based was developed, in part, by the United States Environmental Protection Agency (EPA) through Cooperative Agreement No. CR-831559-01 with the Water Environment Research Foundation (WERF). However, the views expressed in this document are solely those of Malcolm Pirnie Inc. and neither EPA, WERF, WEF, nor NACWA endorses any products or commercial services mentioned in this publication. This report is a publication of WERF, not EPA. Funds awarded under the Cooperative Agreement cited above were not used for editorial services, reproduction, printing, or distribution.

ACKNOWLEDGMENTS

The project team would like to thank the Water Environment Research Foundation (WERF) and co-funding organizations including the Water Environment Federation (WEF), and the National Association of Clean Water Agencies (NACWA). Thanks to Lola Olabode of WERF for managing this project. Thanks to Linda Kelly of WEF and Cynthia Finley of NACWA for providing input to the scope and direction of the project. The team would like to thank representatives at utilities that participated in interviews and provided insightful comments to a draft of this report, including Michael Scahill, Central Contra Costa Sanitary District, California; Bret Linenfelser and Jody Jacobson, City of Boulder, Colorado; Phil Bobel, City of Palo Alto, California; Carie Wilson and Susan Butler, City of Scottsdale, Arizona; Jen Jackson, East Bay Municipal Utilities District, California; Steve Carr, Ann Heil, Don Avila, Debbie Boadway, and Joe Gully, Los Angeles County Sanitation Districts, California; Eleanor Torres and Mike Wehner, Orange County Water District, California; Keith Haas and Michael Kosterman, Racine Water and Wastewater Utilities, Wisconsin; Jose Lopez, Julie Huber, Christine Smith and Mark Elsner, South Florida Water Management District, Florida; and Ron Wildemuth, West Basin Municipal Water District, California. Thanks also to technical reviewers Lois Humphreys of TRG & Associates, Inc. and David Sedlak.

Report Preparation

Principal Investigators:
Rula A. Deeb, Ph.D., BCEEM
Malcolm Pirnie Inc.

Edward G. Means III
Malcolm Pirnie, Inc.

Project Team:
Jozee Adamson
Decision Research

Elisabeth L. Hawley, P.E.
Malcolm Pirnie Inc.

Linda Macpherson
CH2M HILL

Paul Slovic, Ph.D.
Decision Research

Review Committee

William Cairns, Ph.D.
Trojan Technologies

Chris Hornback
National Association of Clean Water Agencies (NACWA)

Linda Kelly
Water Environment Federation (WEF)

Margaret H. Nellor
Nellor Environmental Associates

Elizabeth Toot-Levy
Northeast Ohio Regional Sewer District

Amy Woodis
Metro Wastewater Reclamation District, Denver

Water Environment Research Foundation Staff

Director of Research: Daniel M. Woltering, Ph.D.
Senior Program Manager: Lola Olabode, M.P.H.

ABSTRACT AND BENEFITS

Abstract:

This report provides a state-of-knowledge synthesis of existing work and guidance on effective risk communication practices, public perception, and message effectiveness. Communication principles are applicable to a wide variety of potential health and environmental risks; however, the report is written with a focus on trace organic compounds. Findings are drawn from: 1) a focused literature review of communication materials published in the environmental industry; 2) documents describing risk communication practices in other industries (nuclear energy, chemical manufacturing and the pharmaceutical industry) which culminated in several "lessons learned" that are relevant to trace organic compounds; 3) coding and systematic analysis of approximately 25 recent media articles pertaining to trace organic compounds focused on vocabulary and imagery, key messages, and the articles' likely impact on the public; and 4) interviews with water and wastewater utility representatives to better understand their existing communication and outreach programs, interaction with the public and media and perspectives on communications needs. Principles described in this report can be used to convey a wide variety of messages to help managers better communicate with the media and public. Recommendations for utilities and ideas for future research specific to trace organic compounds are also provided.

Benefits:

- Reviews pertinent documents previously published for utilities on risk communication and provides an analysis of effective communication practices in general, with a focus on the issue of potential risks of trace organics to human health and ecology.
- Identifies and describes case studies where risk communication practices were successful, where they failed, and why (i.e., lessons learned).
- Analyzes the role and impact of the media in framing public opinion on the topic of trace organic compounds.
- Provides recommendations on developing messages, diagnosing risk communication deficiencies, and addressing them in future communications.
- Lists recommendations for future research on the topic of risk communication specifically related to trace organic compounds.

Keywords: Trace organic compounds, PPCPs, emerging contaminants, pharmaceuticals, microconstituents, risk communication, education, perception, public relations, media.

TABLE OF CONTENTS

Note: The Detailed Analysis of Media Coverage is Available by Request

LIST OF TABLES

LIST OF FIGURES

LIST OF ACRONYMS

%	Percent
µg/L	Micrograms per liter
ALDOT	Alabama Department of Transportation
AMWA	Association of Metropolitan Water Agencies
ANEC	American Nuclear Energy Council
AP	Associated Press
AWWA	American Water Works Association
AwwaRF	American Water Works Research Foundation (as of 1/1/2009, the Water Research Foundation)
CAC	California Avocado Commission
CCL	Contaminant Candidate List
CEOs	Chief Executive Officers
COP	Community of Practice
CRC	Cooperative Research Centre for Water Quality and Treatment
CSR	Corporate Social Responsibility
DOE	Department of Energy
GWRC	Global Water Research Coalition
IAP2	International Association for Public Participation
IAT	Issue Area Team
IWAG-TU	Institute for Water Quality, Resources and Waste Management, Vienna University of Technology, Austria
mg/L	Milligrams per liter
MTBE	Methyl Tertiary Butyl Ether
NACWA	National Association of Clean Water Agencies
NCP	National Contingency Plan
NDMA	N-Nitrosodimethylamine
ng/L	Nanograms per liter
NRC	National Research Council
NVC	Nonviolent Communication
NWRI	National Water Research Institute
OCWD	Orange County Water District, California
OEHHA	Office of Environmental Health and Hazard Assessment
PhRMA	Pharmaceuticals Research and Manufacturers of America
PPCPs	Pharmaceuticals and Personal Care Products
PUB	Public Utilities Board
RFP	Request for Proposal
RO	Reverse Osmosis
S&P	Standard & Poor
SARF	Social Amplification of Risk Framework
SDA	Soap and Detergent Association
TPUs	Technical Practice Updates
TZW	Technologiezentrum Wasser, Germany
U.S.	United States

U.S. EPA	United States Environmental Protection Agency
UK	United Kingdom
WEF	Water Environment Federation
WERF	Water Environment Research Foundation
WHO	World Health Organization
WQRA	Water Quality Research Australia
WRC	Water Resources Commission, South Africa
WSAA	Water Services Association of Australia

EXECUTIVE SUMMARY

A variety of trace organic compounds (i.e., part per billion or part per trillion concentrations of steroid hormones, pharmaceuticals, personal care products, disinfection byproducts, surfactants, flame retardants, plasticizers, and other compounds) have been detected in surface waters and municipal wastewater effluent. Although the environmental and human health effects of most of these compounds have not been fully explored, some of these compounds have been shown to cause biological responses at low concentrations, mainly in aquatic organisms (see e.g., WERF report CEC3R07, *Technical Brief: Trace Organic Compounds and Implications for Wastewater Treatment*, 2008). Many of these compounds have likely been present in the environment for decades but have only recently been detected due to analytical developments such as gas chromatography coupled with tandem mass spectrometry and liquid chromatography/mass spectrometry.

The presence of these compounds has raised public concern and spurred research efforts to evaluate the magnitude of this potential environmental problem. Research has shown that people are primarily concerned about risks to human health, followed by the ecological health of streams and rivers, the degradation of groundwater quality, and other concerns. Media coverage can use terms and images that are alarming, increasing public concern, particularly when the public are not presented with enough background or context about water quality. Water and wastewater utilities would like to know how to communicate with the public on this issue, how to identify and alleviate unnecessary fears, how to clear up misconceptions, and how to build public confidence in water quality. Utilities are aware that communication with the public is complex, as technical communications may sound alarmist or be difficult to understand. This communication is constrained by the limited scientific understanding of the human health effects of these compounds; insufficient health effects information makes generalized reassuring statements about safety open to debate. This topic is covered by the news media from time to time, usually as a feature story rather than prompted by any particular news event. Therefore utilities must be prepared at all times to address questions from the public and media on this topic, as illustrated by the public response to a March 2008 Associated Press (AP) article (AP, 2008) on pharmaceuticals in drinking water.

In response to the need to put these issues into perspective, the Water Environment Research Foundation funded research to prepare this report. Though the research does not provide answers to all the questions from the public, the objective of this report is to provide a state-of-knowledge summary and synthesis of existing work that provides guidance on effective risk communication practices, public perception, and message effectiveness with a primary focus on trace organic compounds. Report findings are based on information gleaned from the following activities:

- *Focused literature review.*
    - Review of general risk communication publications, including selected social psychology studies.
    - Review of select publications for utilities pertaining to risk communication on a variety of topics, including trace organic compounds.

- *Literature review of other industries.*
  - Review of publications for other industries summarizing risk communication principles and lessons learned. Other types of industries have also faced challenges in risk communications. Consequently, several articles published on the topic of risk communication efforts with the public by chemical manufacturing companies, the nuclear energy industry, and pharmaceutical manufacturers were also reviewed.
- *Media analysis and input from utility teaming partners.*
  - Review and analysis of media articles on trace organics. A selection of recently published media articles about trace organic compounds were reviewed and analyzed for choice of vocabulary, imagery, key messages, and other aspects. The objective of this work was to summarize how the issue of trace organics is presented by the media and to gauge the likely impact on public understanding and emotions.
  - Interviews with ten different utility teaming partners (water and/or wastewater utilities). Representatives from each utility participated in a telephone interview to answer questions about their experiences and outreach programs interacting with the public and the media about trace organic compounds.

Based on this research, a general framework for communications was developed. This framework was prepared to illustrate steps in the communications process for utilities wishing to improve their existing communication program on trace organic compounds. In addition, based on this research, recommendations for utilities and ideas for future research on risk communication related to trace organic compounds were identified.

ES.1 Focused Literature Review

Social psychology and risk perception research can provide a helpful context for better understanding and anticipating people's mindsets, decisions, and reactions to messages. Research has shown that people make decisions using a set of intuitions gained from other experiences and often rely on their affective or emotional response, particularly in unfamiliar situations. Word choice and imagery can therefore strongly influence peoples' affective response, shaping their initial perceptions of an activity or technology. People may respond to certain risks with an excessive negative reaction based on their past experiences and/or fears associated with a sign or marker that they recognize from another context. Finally, the public's broader concept of risk includes factors beyond those considered in scientific risk assessments (e.g., individual experiences and culture, dread, controllability, potential for catastrophe, risk to future generations, power inequity, economics, and politics).

Research that was reviewed described the important role that trust plays in facilitating smooth and successful communication. Research has shown that trust is easier to destroy than to create, i.e., trust-decreasing events have more impact than trust-increasing events. In addition, people find sources of bad news to be more credible than sources of good news. Suggestions for building trust include consistently "walking the talk," (i.e., behavior that is consistent with key messages), identifying and meeting stakeholders' basic expectations, valuing the decision-making process, collaborating with reputable third parties or programs, appearing knowledgeable and experienced (i.e., providing educational information), being honest and transparent, and displaying concern and care. Some of these suggestions go beyond the style of communication and address the underlying activities and behaviors of an organization.

Given this context, what are people's opinions and beliefs about environmental issues, including water resources and, in particular, trace organic compounds? A number of studies have been conducted to gauge public opinion about specific environmental topics, including water conservation and shortages, tap water, recycled water, biosolids, and trace organic compounds. Only one study was specific to trace organic compounds (termed "emerging contaminants" in the study) (AwwaRF, 2004). Results indicated that people did not understand the term "emerging contaminants," would like to know what contaminants had been found, what the hazards were, what their local utility planned to do about the situation, and what people could do to protect themselves (AwwaRF, 2004). The lack of studies specific to people's perceptions on trace organic compounds was noted and is an area recommended for future research.

Studies have shown that people's opinions and ideas about contamination, water resources, and other environmental issues are amplified and directed by media coverage. Most people get their information from television and rate television news as highly believable. Although the believability of local news channels declined from 1985 to 2004, local and network television news were both rated more believable than newspapers over this time period (Perras, 2005). Previous studies have pointed out that national television coverage and the increasing use of wire copy reports distributed by large organizations can increase the impact of a single news story. Prior studies have recommended preparing simplified messages focused on the key issues that can be delivered quickly to the media in a format that meets their needs, including television interviews (a 10-12 word sound bite), quotes (one to three lines), and press releases (facts and keys messages). More details can be provided via a website or upon request.

Prior studies have also summarized the process of preparing a communication plan. Such a plan demonstrates the organizations' commitment to risk communication and walks the authors through a thought process about communication principles. Several frameworks have been published to summarize the process and content of a communication plan (see for example, the summary listed in Table 3-1). The cyclic nature of developing and refining a communication plan over time is illustrated in the communication framework presented in Section 5.0. Successful implementation of the communication plan is described in Section 3.5.3 of this report. Examples of communication vision statements and communication plans were not encountered during the literature review; this process may not be well-documented with examples for utilities to follow.

Communication plan implementation may include educational activities that further the public understanding of water (as a mixture of many different compounds, as well as a background on basic treatment processes) and provide perspective without trivializing the issue of trace organic compounds. Communication plan implementation will likely involve giving presentations at public meetings and to the media. Tips for effective spokespersons are summarized in Section 3.5.3.5. Spokespersons appear more trustworthy when they display knowledge/expertise, empathy and care, honesty, and commitment to the project. ASTDR (2004) lists pitfalls for spokespersons to avoid, including the use of jargon, repeating negative allegations, using body language and other nonverbal messages that are inconsistent with the main message, speculating, or making "off the record" remarks. Most importantly, spokespersons need to focus on key messages throughout formal presentations and question and answer sessions.

As described previously, documents are not currently available that identify for utilities what the key messages "should be" in communications about trace organic compounds. Certainly, there are references that summarize the current state of knowledge on trace organic compounds, including potential human and environmental health effects, occurrence, chemistry, predicted or measured treatability in conventional wastewater treatment systems and other topics. (See for example, WERF report CEC3R07, *Technical Brief: Trace Organic Compounds and Implications for Wastewater Treatment,* 2008.) However, the distillation of key messages for the public has not yet been conducted. There is no consensus on what the key messages are, and key messages are likely to vary based on the topic at hand (e.g., drinking water, wastewater) and the utility's perspective on the issue. The development of key messages and testing of their impact was identified in Section 6.0 as an area of needed research.

In the absence of clear answers that are scientifically based, communication efforts, trust, and "how to say it" become particularly important, as well as continued efforts to provide basic water education. Scientific research and communications will continue in parallel and key messages will change over time. One of the key difficulties facing utilities on the topic of trace organic compounds is communicating in the face of uncertainty. As demonstrated by the literature review of other industries, this has been encountered before by other industries but has not stopped communication efforts.

ES.2 Literature Review of Other Industries

A selection of articles on risk communication in other industries (nuclear power, chemical manufacturing and pharmaceuticals production) were reviewed to identify and illustrate effective communication principles that may be applicable to the issue of trace organic compounds and other environmental issues. Both the nuclear and chemical industries have faced serious public opposition in the past. The pharmaceuticals industry may be perceived as less of a threat to public health because 1) the benefits of medications are perceived to be strong and 2) the industry takes communications and marketing (product image) seriously. All three industries address trace levels of contaminants that the public is inadvertently exposed to. Findings underscore several of the same concepts presented in risk communication documents for water and wastewater utilities. Six lessons learned were gathered from this literature:

Pay attention to risk communication. Organizations who do not conduct risk communication allow their reputations to be defined by others. Information gaps will be filled with peoples' worst fears. Communication in the face of uncertainty is therefore particularly important. In the 1960s and 1970s, Dow Chemical Company's name became synonymous with environmental and health controversies. Dow did not pay attention to public opinion until changes in senior management style in the 1980s, which led to employee opinion surveys and a shift in corporate culture, outreach and re-branding (Long, 2001). The importance of consistent risk communication was highlighted by Peter Sandman (2004) in his review of the Three Mile Island incident of nuclear power plant malfunction in 1979. After the incident, communication was often garbled with multiple spokespersons presenting conflicting information to the public and the press.

Build trust. Trust can be built by emphasizing transparency, honesty, personal control, and community participation/oversight. Providing clear, open, and honest information/education enhances trust. The link between trust and these principles has been demonstrated by several surveys (Slovic, 2007a; 2007b), media incidents such as Johnson & Johnson's response to media attention in the 1980s (van Ryan, 2003), the success of community participation in the Superfund cleanup process (van Ryan, 2003) and analysis of events at Three Mile Island (Sandman, 2004). Trust is particularly important when communicating about accidents and other uncertain topics.

Poll perceptions before designing a public relations campaign and during the campaign. This practice can save money, identify and focus the campaign on key messages, and gauge or predict the campaign's effectiveness. An example of a failed advertising campaign is described by Flynn et al. (1993). The campaign was designed to generate public support of a high-level nuclear-waste repository in Yucca Mountain, Nevada. However, an opinion survey conducted after the campaign was launched indicated that 29% of viewers actually decreased their support for the project after watching the advisements, i.e., the campaign was counter-productive.

Emphasize benefits. Although risk and benefit are typically positively correlated in society, people tend to classify risks and benefits as being negatively correlated (Slovic, 1999). This perceived inverse relationship is attributed to people's affective evaluations when making risk and benefit judgments. That is, if people have an overall unfavorable reaction to a technology or activity, they will rate it as low-benefit and high-risk. If they have a favorable reaction, they will rate it as high-benefit and low-risk (Slovic, 1999). Emphasizing benefits and value (perhaps through information and education), thereby increasing positive associations with a technology or activity, will result in a lower perceived risk. In the face of uncertainty, lower perceived risk will not be as alarming.

Adopt inclusive decision-making processes. People value personal choice and individual autonomy; they are therefore more comfortable accepting risks that are voluntary (Covello, 1991). People also prefer having some decision-making power when a technology or activity involves risk, particularly when risks are not fully understood. Examples of successfully involving the public as decision-makers in technical situations are outlined by Drew et al. (2003) and van Ryan (2003).

Recognize the role and implications of stigma. The concept of stigma may be useful in understanding situations, technologies, or activities that are perceived as unduly dangerous. The word "stigma" implies something benign or beneficial that suddenly becomes menacing – water contamination may be an example of such a situation. Media reporting can be an impetus for the creation of stigma by reporting a situation as hazardous, involuntary, and dreaded, and as a result, sending a strong signal of risk (Gregory et al., 1995). Stigma can be experienced as a sense of contagion or contamination, where some marked substance threatens the body regardless of the dose. Scientists often think of stigma as similar to superstition or irrationality, and tend to minimize its public relevance. When working with the public, practitioners and scientists need to therefore be sensitive to potential pitfalls associated with trivializing stigma and contagion concepts (Walker, 2001). The images and ideas associated with stigma are a potent force and may only be addressed through a different framing of the issue. Information, education, and trust are important considerations. Kunreuther and Slovic (2001) describe strategies for dealing with stigma, which are listed in Table 3-2.

These six lessons learned can be used to address risk communication on trace organic compounds and other environmental issues. While many industries faced the need to communicate in the face of uncertainty (e.g., nuclear industry), uncertainty did not stop them from communicating with the public about the issue (e.g., Three Mile Island incident and other accidents) and the underlying communication concepts and trust-building principles were particularly useful during these times.

ES.3 Media Analysis and Input from Utility Teaming Partners

As described previously, media coverage can help shape public opinion on environmental issues, requiring a response from the water industry. The researchers analyzed 25 recently published media articles and one cartoon image that addresses the issue of trace organic compounds. Several of the articles were published after the March 10, 2008 AP coverage of pharmaceuticals in drinking water. The selection of media articles is neither comprehensive nor representative of nationwide coverage of trace organic compounds. Nevertheless, some clear messages in these articles are worth considering when communicating with the public and media.

A uniform logic was applied to each article to standardize the impressions that each article gave the reader. A set of coding categories was created based on the theoretical model of "*social amplification of risk framework*" (Kasperson et al., 2003), which focuses on key messages or "signals" emanating from a risk incident or news report of the incident, which can intensify public concern. The overall affective (positive or negative) tone of the article and 14 other coding categories were evaluated:

- Overall affective tone
- Failure to disclose information
- Pharmaceutical names listed
- Other substances listed
- Scope of contamination
- Concentration and amount of exposure
- Risk assessments
- Limits of control technology
- Observable harm
- Culpability
- Regulation
- Emotional and inflammatory images and terms
- Sensitive populations
- Public response
- Solutions
- Key messages

The overall affective tone of the 26 articles was negative. Approximately one-third of the articles were deemed likely to trigger a strong negative reaction. Only three articles were positive, three were mixed or balanced, and two were factual. About one-third of the articles reported a failure to disclose information about chemicals in water. Typically, the articles referred to pharmaceuticals and other compounds using common names. Almost all media articles gave some indication of the scope of contamination in terms of the geographic location of affected water supplied and the number of people exposed. (Five articles stated that millions of people were exposed). Concentrations were typically given in perspective using analogies or were just referred to as "tiny," "low" or "miniscule." About 80% of the articles assessed risk in some way, typically in the form of assertions or implications that the water was safe or low in risk. There were also statements of concern over risks to people or the environment. Several articles presented a mix of assertions about safety and others focused on the uncertainty of health risks. About one-third of the articles mentioned that treatment plants were challenged to remove these compounds from water, while a few articles indicated that compounds passed through treatment untouched or that only trace amounts of the compounds remained after treatment. Nearly all articles mentioned specific observed harm to humans or a range of aquatic organisms, including feminization of male fish. About half of the articles indicated culpability but most put the blame on the public. About half mentioned a lack of regulation or standards, several calling out U.S. EPA as not adequately regulating chemicals in water or utilities for not testing drinking water supplies. Only two articles mentioned greater sensitivity to chemicals among certain individuals. Four articles contained material on potential or actual public responses. Half of the articles mentioned possible solutions to the problem with state-of-the-art treatment technologies. Key messages included that chemicals in waterways were a new risk to worry about, that experts appear worried and uncertain about the degree of risk, that risks may be unfair or uncontrollable, not well-regulated, and may portend long-term consequences to human health.

Overall, the articles portray the detection of trace organic compounds as an emerging problem. They point to findings of sex-related changes in fish that create an unsettling feeling and send a signal of potential serious consequences to humans. They indicate that regulatory agencies are not doing an adequate job. Consistent with other risk coverage that generated massive public concern, media coverage indicates that trace organic compounds are uncontrollable, create involuntary exposure with dread consequences, and are poorly regulated by authorities. In the selection of articles reviewed, few attempts were made by the media to educate the public about broader issues related to water or to inform the public about modern water treatment techniques to make water reuse safe and beneficial.

The impact of media coverage on trace organic compounds was discussed with representatives from 10 different utilities of differing sizes in various locations (California, Arizona, Colorado, Wisconsin, and Florida) who were identified and approached during the proposal phase as participants pledging in-kind contributions. Some states were experiencing water shortages, the majority of utilities interviewed practiced some sort of water reuse, and three had advanced wastewater treatment systems. Results are not representative of nationwide perspectives and are not statistically significant. However, these utilities provide examples of the level of public involvement, types of communication programs, and utility responses to media coverage. Uniform logic was applied to the interviews in an attempt to standardize input from different utility organizations, with a set of 11 questions asked of each representative (these are listed in Section 4.3.2).

Interview responses indicated that the media is taking the lead on how trace organic compounds are portrayed and understood by the public, which shapes the communication challenge for utilities. Media coverage has had a dramatic effect on public interest by spotlighting the issue for the public. For example, terminology to describe the issue is often chosen by journalists, rather than initiated by the water/wastewater industry. Some utility representatives interviewed thought that media sensationalism and a lack of balanced and informed coverage was due to market pressures and the trend of hiring less experienced journalists. Working with the media proactively was mentioned as one solution for enhancing media understanding. Utilities would like to be proactive and are concerned about how to respond to media attention. Only one utility proactively told the public about an unregulated compound found in the water supply. This utility described their experience as positive. All others have found that they have either provided information to the public or media reactively or have not experienced public interest on this topic.

Among utilities, there are different approaches to addressing the issue of trace organic compounds and varying levels of importance of this issue within the organization. On one end of the spectrum, utilities believe that the issue is very important to the organization and should be characterized as an emerging issue or as a "concern." For at least one utility, the issue was not a concern at all. Utilities that are communicating with the public on this topic have both communication specialists and technical experts responding to the public, depending on the type of question. There has been little formal tracking of the number of requests and the impact of utility communications on the level of public concern. The majority of utilities interviewed currently use or are planning pharmaceutical take-back programs so that people can drop off unused medications rather than flushing them down the drain or disposing of them in household trash. The state of California is in the process of setting up a statewide pharmaceutical take-back program, a process that may be time-consuming due to significant statutory impediments. (For example, authorized officers of the law may be required to oversee collections.) Other utilities are practicing or advocating product substitution. Most utilities viewed these programs as a way for the public to be a part of the solution, to partially reduce concentrations of pharmaceuticals entering wastewater treatment facilities, and to provide an opportunity to interact with the public about the topic, answer their questions about the issue, and provide them with additional information. Utilities highlighted the need to partner with credible organizations, perhaps outside of the water industry, to help the public understand the issue, including potential health impacts, as water and wastewater experts may be perceived as having a conflict of interest.

Utilities noted that the public does not have the necessary background to understand the issues in context, and they may not understand the issue of detection, the advancement of analytical techniques, or the meaning of different concentrations. Public concern seems to be focused on human health impacts more than effects on aquatic life and other ecological species. People's concerns about risks to humans are likely heightened by studies and images of reproductive effects in aquatic species, indicating that human health concerns may be feared but unsubstantiated. Utilities noted the interest in involving organizations without a financial interest in the outcome (e.g., public health agencies) to better frame public discourse on the topic and enhance public understanding.

ES.4 Communications Framework

A significant amount of information has been published about effective risk communication practices. This report presents a summary of the information that will be most useful for utilities, organized into several basic steps: 1) developing a communication vision, 2) preparing a communication plan, and 3) implementing that plan effectively. Developing a communication vision is similar to the process of brand development, a well-established area of market research. Like branding, the objective of a communication vision is to identify a few key messages that can be communicated consistently in an easily recognizable way.

Based on the literature review, a general framework for communications was developed (Section 5.0). Such a framework was prepared to illustrate the steps in the communications process for utilities who wished to improve their existing communication program on trace organic compounds. This general communication framework lends perspective to a utility's communications with the public and media by showing the types of "behind the scenes" work that can improve the effectiveness of communication. These steps have been summarized as 1) understanding stakeholders, 2) developing a communications plan, 3) aligning programs with that plan, 4) designing and disseminating information (followed by monitoring results to once again better understand stakeholders). Of these steps, the majority of the work is conducted "behind the scenes" to make sure that the communication style and content will be effective in achieving the organization's communication goals. The majority of the steps also provide opportunities for the utility to build or strengthen public trust. The primary purpose of the communications framework is to place actual communications (e.g., brochures, public meetings, tours, webinars) in perspective of the greater communications process. Another potential benefit of the framework is to help utilities think through the communications process so that they can identify areas of opportunities for strengthening their communication programs on trace organic compounds.

ES.5 Summary and Recommendations

Based on the research summarized previously, recommendations for utilities and ideas for future research on risk communication related to trace organic compounds were identified (Section 6.0). The top five key recommendations for utilities include the following:

1. *Talk with stakeholders*. Gain a perspective on the public's values, interests, concerns, and priorities; their understanding of trace organic compounds and other water quality issues; their attitudes towards the utility; trusted information sources; suggestions for addressing their concerns; and other topics. This information can be used to fill gaps in information or perception and can fuel changes in key messages, programs and practices.

2. *Understand the media's needs and tailor communications accordingly*. Based on the researchers' review of media articles and utility interviews, there is opportunity for utilities to provide their facts and perspective on trace organic compounds through the media. To use this communications channel effectively, prepare succinct statements of key messages for broadcasts, press quotes, and press releases, with supporting information available online or on request.

3. *Collaborate with other organizations*. Based on the literature review and interviews with utility teaming partners, collaboration is recommended to increase trust and strengthen support of key messages that the public receives on trace organic compounds. This collaboration could include local universities or health professionals.

4. *Involve staff in the communication process*. Based on the literature review and interviews with utility teaming partners, staff contribute significantly to the face of the organization and therefore need to be involved in the communication planning, message development, and dissemination. Training programs and other informal interaction with staff on communication-related topics are therefore both diagnostic (e.g., how is the utility viewed?) and remedial (e.g., how can staff address specific public comments and questions?).

5. *Define the utility's communication vision on trace organic compounds*. Define the utility's key values and overall objectives of the communication program, as described in Section 3.5.1. This is the first step in designing a communication plan on trace organic compounds and can influence the development of key messages, partnerships, program changes, and other communication plan elements.

   Several other recommendations for utilities are summarized in Section 6.1.

In addition to recommendations for utilities, a number of recommendations for future research were compiled (see Section 6.2). The top five recommendations for future research include the following:

1. *Conduct research and develop guidance to inform utilities' messages on trace organic compounds based on public understanding and response to a variety of issues.* Based on the lack of consensus on messages specific to trace organics highlighted by the literature review, this is the primary recommendation for future research. Priority topics to be addressed by this research include effective communication methods for describing detection and concentrations, development of consensus methods regarding potential human health concerns, and the effect of different partial solutions (e.g., advanced treatment, pharmaceutical take-back programs) on the public's perception of risk.

2. *Survey utilities to identify common goals and explore areas for inter-organization collaboration.* In order to leverage resources and improve message consistency, utilities may want to establish an interagency educational outreach program to develop common messages and communicate these with the public. Future research could analyze the similarities and differences between utilities' key values, messages, and priorities to explore the viability of establishing a national spokesperson or organization, identify the amount of common ground among utilities, and identify several core messages that could be the basis of an informational campaign or risk communication strategy for the group.

3. *Develop tested communication tools and templates for utilities.* Sample press releases, brochures about trace organic compounds, a "how-to" guide for responding to and seeking out the media, an in-house training guide for technical staff, a summary of key articles and ongoing research, and a brief (five-minute) video about trace organic

compounds are just some examples of the tools and templates that utilities may need. Such tools would require testing to ensure that they were beneficial for utilities and the public.

4. *Conduct a detailed analysis of the impacts of media coverage and develop recommendations for utilities to interact with the media.* As described in this report, media coverage can amplify and direct public attention. Research could expand on the preliminary analysis of media coverage of trace organic compounds presented in this report and be used to develop guidance for utilities in communicating with the media. Based on interviews with utility partners, many utilities want to be proactive in communicating with the media but are concerned that the information will be presented without context, increasing public alarm unnecessarily.

5. *Establish an online education/media center for journalists and the general public.* Based on the researcher's analysis of media articles and interviews with utility teaming partners, there is a need to provide educational materials to the public and to the popular media that are easily understandable and accurately reflect scientific information about trace organic compounds. Assembling this material on a website would counter the shallow coverage that these topics are given in the popular press and provide the public with more information to inform their opinions and decisions about water resources and contamination. To make the content more appealing to a broader audience, website text could be accompanied by images, webcasts, and/or podcasts.

Several other recommendations are summarized in Section 6.2.

CHAPTER 1.0

INTRODUCTION

Recent improvements in analytical methods have revealed myriad chemicals in wastewater effluent and receiving waters at low concentrations (often sub part per billion levels). Many of these same compounds have been detected in surface waters, groundwater, and drinking water systems. Despite their low concentrations, some of these compounds may cause biological responses. For example, consumption of water contaminated with 3 nanograms per liter (ng/L) of N-nitrosodimethylamine (NDMA) has been estimated to pose a one-in-a-million lifetime excess cancer risk (Office of Environmental Health and Hazard Assessment (OEHHA), 2006). Biologically significant effects associated with ethinyl estradiol (EE2) in experimental lake settings over a seven-year study period have been reported, despite low concentrations in the range of 10 to 150 ng/L total estrogenicity (expressed as 17β-estradiol (E2) equivalents) (Kidd et al., 2007). Context for aquatic effects is provided in the WERF report *Technical Brief: Trace Organic Compounds and Implications for Wastewater Treatment* (WERF, 2008). The detection of a large number of wastewater-derived trace organic compounds and documented risks posed by select contaminants have raised media interest, public concern, and spurred research efforts to try to evaluate and communicate the actual magnitude of the problem in a balanced and informative manner.

Consistent with a previous WERF publication, the term "trace organic compounds" includes organic compounds that have been detected in water and wastewater samples in trace amounts (low part per billion concentrations or less). Examples include, but are not limited to, steroid hormones, pharmaceuticals, personal care products, disinfection byproducts, surfactants, flame retardants, and plasticizers. A list of trace organic compounds is presented in Appendix A of *Technical Brief: Trace Organic Compounds and Implication for Wastewater Treatment* (WERF, 2008).

1.1 The Need to Communicate Information about the Human and Environmental Health Effects of Trace Organic Compounds

There is a need to communicate information about trace levels of organic contaminants. However, because there are a number of different compounds, a variety of potential human health and environmental effects or lack of information about effects, and challenging analytical methods, it is difficult to scientifically support generalized conclusions about the effects of trace organic compounds. Many research studies are still ongoing (e.g., the development of acceptable daily intake values or threshold concentrations) and the field of knowledge is quickly evolving, making it difficult to draw general conclusions about the topic. A recent overview of the key issues associated with trace organic compounds was published by WERF (2008), including an introduction to the problem, current analytical methods, removal during wastewater treatment, potential human health and ecological effects, and the response from different regulatory agencies. A significant number of trace organic compounds are listed on the U.S. Environmental Protection Agency (U.S. EPA) Contaminant Candidate List (CCL 3) or are slowly being evaluated under the Endocrine Disruptor Screening Program; other compounds have not yet been addressed by any regulatory program. A final determination of whether or not any of these compounds will be regulated under the Safe Drinking Water Act or the Clean Water Act is unlikely prior to the collection of additional data on occurrence, toxicity, and treatability.

1.1.1 Public Concerns

Meanwhile, public health agencies, utilities, industry, and other organizations are faced with the task of communicating with the general public and with the news media about the topic in the absence of information without causing undue fear or alarm. Many people are concerned about the safety and healthfulness of their drinking water, the ecological health of streams and rivers, the degradation of natural resources, and other effects. In addition, people are typically concerned when today's practices may cause future adverse impacts. If so, they would like to know what the impacts might be and what options may be available to correct the situation. People may also want to know who, if anyone, is paying attention to the issue, who is in charge of making decisions and implementing solutions, and whether or not these organizations or individuals can be trusted (e.g., are special interests involved?). Based on the answers to these questions, members of the public will form a general opinion or impression about the topic, decide which information source(s) to trust, and may make personal choices in an attempt to reduce or manage risks (e.g., using a point of use treatment device or drinking bottled water).

1.1.2 Communication Goals

The goals of communicating with the public (either directly or indirectly, through the media), regulators, and other entities may differ from one organization to another and may vary based on the type of organization (public health department, water treatment technology vendor, water or wastewater utility, industry, environmental advocate). Examples of potential motivations include the following:

- To protect the organization's reputation and/or business interests.
- To satisfy their public duty/responsibility or as a public service, sharing what technical information they have with the public and perhaps explaining how they themselves have personally responded to the issue. For example, utility employees might describe whether or not they filter their drinking water at home or dispose of excess prescription drugs through a take-back program.
- To promote or maintain public trust.
- To initiate dialogue and build a relationship with the public about water quality issues. This may be risky as it is not always easy to predict the reaction of the public – does communication foster trust via transparency or instill fear?
- To protect legal interests or abide by the law, avoiding future lawsuits.
- To support a special cause or relay information that may be of interest to their constituency. For example, Catholic groups have published news articles about pharmaceutical compounds in the environment as another reason for women not to use birth control pills (Frazier O'Brien, 2008).
- To stay informed about upcoming regulations, monitoring, and treatment requirements and to have a dialogue with regulators about the feasibility and effectiveness of different approaches.

Different communication goals will likely influence the formulation of key messages, target audience, method of disseminating information, desired spokesperson qualities, and choice of other communication tools and techniques.

1.1.3 Communication Impacts

Sometimes, technical communications may sound alarmist or difficult to understand. The idea of invisible drinking water contaminants that may cause cancer or disrupt the endocrine system is scary. Without a science background (and sometimes even with), people may lack the framework to put the story into context and may misunderstand the information they are given, unless it is reported in context. They may not understand the concept of concentration. Or, people may not understand the difference between milligrams per liter (mg/L), micrograms per liter (μg/L) or ng/L. Reporting a concentration as 0.00001 mg/L NDMA will seem much smaller to them than 10 ng/L. When scientists report numbers to the public, they may not provide adequate perspective or context for interpreting the concentration. It may be simpler to avoid numbers and only provide comparisons for a sense of scale. Photographs and graphics provided along with text are also important in peoples' minds and strongly contribute to their overall impressions of the communication. The frequency of communication may influence peoples' interpretation as well; people may interpret communications as more serious if they are rare.

1.1.4 Media Coverage

Organizations that do not actively communicate about an issue are effectively silencing their own voice while other voices continue to reach the public, shaping their impressions and opinions, and becoming the preferred source of information. Many people get their information through the media, particularly television news, which will likely be dramatized, polarized or presented briefly without much context, due to the need to quickly interest and retain viewers by presenting a variety of stories. Internet-based news has trained people, particularly the younger generation, to expect news and information instantaneously. The public may think of news stories as presented on behalf of the public, as the media can act as a public watch-dog, instead of seeing news stories as alarmist. The news media's choice of words and graphics indicates that most news coverage on trace organic compounds is typically negative, as described in Section 4.0 of this report.

In March 2008, the Associated Press (AP) released a series of three articles describing pharmaceuticals in drinking water supplies. The AP article was published with different headlines in 746 different newspapers, including the Washington Post, New York Times, USA Today, Canadian Press, and the Guardian United Kingdom (UK). The article presented results of an AP special investigation of pharmaceutical concentrations in various drinking water supplies throughout the nation. Because of the extensive media coverage, the impact of this series of articles was far-reaching, with utilities receiving calls nationwide and organizations such as the Water Environment Federation (WEF) and American Water Works Association (AWWA) preparing and issuing press releases and talking points for utilities. The AP had worked on the series since fall 2007 and had contacted AWWA, U.S. EPA, and over 100 different utilities so utility organizations were given some warning before the story was published. Members of AWWA received an email advisory on the Thursday before the article was released, urging utilities to contact the local media before the article was published. The AWWA Public Affairs office also phoned representatives at each of the utilities mentioned by name in the article to let them know about the upcoming article release date.

Coverage of the issue in the press generated telephone calls from the public to local utilities. Utilities prepared talking points and discussed various aspects of the issue with members of their public. The experience confirmed for many utilities that they need to be prepared to quickly respond to questions from the public on this issue. It also highlighted the need for and the importance of ongoing communication, despite having incomplete research on the topic. These organizations face the challenge of fairly and honestly presenting available information to the public without taking it out of context too much so that it is misinterpreted or unduly causes fear and alarm. In addition to, and perhaps exacerbating this challenge, is the challenge of deciding what to communicate, given the relative newness of the topic and the bewildering variety of trace organic compounds that may be a potential concern. A number of different organizations are already funding research on this topic, as described in Section 1.2.

1.2 Existing Communications Efforts and Programs

1.2.1 Water Environment Research Foundation

The Water Environment Research Foundation (WERF) addresses risk communication related to trace organic compounds through their research and knowledge area on trace organics. WERF resources on trace organics are available on their website (www.werf.org). In May 2007, WERF hosted a workshop on trace organic compounds to collaboratively prepare a roadmap of future research plans and directions (WERF project 06-CEC-1W). Representatives from approximately 30 different organizations from 10 different countries attended the workshop and presented an overview of their research plans. Workshop participants then worked in groups to review and summarize research plans in each topic area (e.g., treatment, fate and transport, communication), identify overlapping interests, and highlight potential areas for future collaboration. During workshop discussions on communication, it was recognized that future research plans fit under one or more of four categories: information development, methods, messages, and partnership development. Multiple organizations identified this area as a priority for future research. Workshop participants recognized that research on communication topics would ideally be conducted concurrently with research on other aspects of trace organic compounds. That way, research could inform communications and the feedback from communications research could also drive research needs and directions.

After the workshop, WERF funded this research (WERF project CEC2R08C) as the first phase of communications research under the umbrella of the trace organics program area. ***The objective of this report is to provide WERF subscribers and other stakeholders with a state-of-knowledge report that is a synthesis of existing work and provides guidance on effective risk communication practices, public perception, and message effectiveness with a primary focus on trace organic compounds.*** Findings are drawn from a review and synthesis of existing documents pertaining to the topic, recent media articles pertaining to trace organic compounds, and documents describing risk communication techniques and practices in other industries. This report will present key concepts that are applicable to a wide variety of potential health and environmental risks with a focus on trace organic compounds.

On its website, WERF also has published a number of materials that may be helpful in preparing for public communication on trace organic compounds, including question and answer sheets, fact sheets, presentations, technical briefs and reports, and links to other organizations' ongoing technical research programs and latest research findings (www.werf.org).

1.2.2 Research Organizations and Industry Associations

Communication efforts on trace organic compounds have also been addressed by WEF. WEF's Community of Practice (COP) on microconstituents has focused on the impact of language and imagery. WEF has published several Technical Practice Updates (TPUs) to aid in communication efforts. WEF has also compiled a glossary of terms, a position statement and other resources on trace organic compounds that are freely available to WEF members (www.wef.org).

AWWA has actively addressed the issue of risk communication through a series of webcasts, conferences, and journal publications. The American Water Works Association Research Foundation (AwwaRF) has funded or is funding over 30 projects associated with trace organic compounds, several of which focus on communication aspects, including *Communicating the Value of Water* (AwwaRF, 2008a) and *Risk Communication for Emerging Contaminants* (AwwaRF, 2004).

The WateReuse Association has also highlighted the importance of risk communication during technical conference sessions. The WateReuse Foundation has funded several projects focused on risk communication, including *Best Practices for Developing Indirect Potable Reuse Projects: Phase 1 Report* (WateReuse Foundation, 2004a) and *Marketing Nonpotable Recycled Water: A Guidebook for Successful Public Outreach & Customer Marketing* (WateReuse Foundation, 2006). They recently issued a request for proposals for a project "Talking about Water: Vocabulary and Images that Support Informed Decisions about Water Recycling and Desalination" (WateReuse Foundation, 2008).

The National Water Research Institute (NWRI) is a public-private partnership formed to promote the protection, maintenance, and restoration of water supplies through the development of cooperative research work. NWRI has co-funded a number of research studies on trace organic compounds in partnership with AwwaRF, WERF, WateReuse Foundation, the U.S. Bureau of Reclamation, Southern California Salinity Coalition, and the California Methyl-Tertiary Butyl Ether (MTBE) Research Partnership. NWRI has funded research on public perception and participation in water reuse initiatives, an issue that shares common ground with trace organic compounds communications. NWRI has also expressed interest in funding projects on the policy, regulatory and public acceptance issues of trace organics in water supplies.

The National Association of Clean Water Agencies (NACWA) has an emerging contaminants workgroup to track, coordinate, and disseminate technical, regulatory, and policy updates. NACWA also drafts and distributes advocacy-focused issue papers to influence the direction of regulatory actions and scientific studies, to assist members in their efforts to improve communication, and to effectively engage their constituents. For example, in response to the recent AP articles, NACWA published a press release summarizing NACWA's activities and progress in establishing a national strategy to address the complex issue of pharmaceuticals and personal care products (PPCPs) in the aquatic environment, partnership efforts with the Product

Stewardship Institute to develop a comprehensive national program for managing the disposal of unused pharmaceuticals, support of local pharmaceutical take-back programs, and funding of research projects to investigate the potential aquatic and human health impacts as well as long-term exposure issues.

Industry associations who have funded communications-related efforts on trace organic compounds include the Pharmaceuticals Research and Manufacturers of America (PhRMA) and the Soap and Detergent Association (SDA). PhRMA has established the Pharmaceuticals in the Environment Task Force to address the concern that human health and aquatic life impacts could result from environmental exposure to pharmaceuticals. PhRMA is interested in developing communications and conducting outreach activities. SDA is a non-profit trade association representing manufacturers of household, industrial, and institutional cleaning products, manufacturers of their ingredients and finished packaging, as well as oleochemical producers and chemical distributors to the cleaning product industry. SDA members produce more than 90% of the cleaning products marketed in the U.S. SDA education efforts are directed at understanding the issues and needs of consumers and addressing them by providing materials that encourage the safe, effective, and responsible use of household cleaning products. They have conducted a number of studies on surfactants and volatile organic compounds.

1.2.3 Water and Wastewater Utilities

Most utilities have well-established communications programs. Several utilities have integrated communications on trace organic compounds into their programs, particularly those practicing water recycling. For example, Orange County Water District (OCWD) participated in the 2007 WERF workshop and shared their past, ongoing, and future research plans with the group. OCWD manages the groundwater basin under north and central Orange County that supplies water to more than 20 cities and water agencies, serving more than 2 million Orange County residents. Since its formation in 1933, OCWD has replenished and maintained the groundwater basin while more than doubling the basin's annual yield. Most of their research has been related to treatment and studies. OCWD is leading the way in purification of wastewater for reuse to provide a reliable, new, drought-proof source of water. The Groundwater Replenishment System, a joint project sponsored with Orange County Sanitation District, will produce enough near-distilled quality water for 144,000 families. Secondary treated wastewater is 100% treated by microfiltration, reverse osmosis, and advanced oxidation with ultraviolet light and hydrogen peroxide. Highly-treated product water is used as source water for a seawater injection barrier at the coast and is also piped 13 miles inland for percolation into recharge basins to replenish the groundwater.

OCWD has prepared a number of outreach materials related to the Groundwater Replenishment System that address trace organic compounds, conducted tours of the facility, and has developed partnerships to facilitate communication. OCWD has worked with the Santa Ana Watershed Project Authority, local physicians, regulators, community groups, and religious groups regarding trace organic compounds. OCWD has given significant thought to the diversity in their community, the need to communicate through a variety of different methods, and the need to communicate information that is important to different groups in the community (e.g., consulting with the local imam to make sure that recycled water would be halal (i.e., permitted under Islam, particularly in reference to food and drink).

1.2.4 International Research

There is significant international interest in risk communication on trace organic compounds. Singapore PUB is internationally recognized for the communications program accompanying their state-of-the-art recycled water (NEWater) facility. Other leaders in research include Water Quality Research Australia Limited (WQRA), formerly known as the Cooperative Research Centre for Water Quality and Treatment (CRC Australia), a national not-for-profit scientific research institution with past, ongoing, and future research related to knowledge transfer of research results, educational activities, and public outreach. Other international research organizations who attended the 2007 WERF workshop and shared plans or expressed interest in conducting future research on communications-related topics on trace organic compounds include the Global Water Research Coalition (GWRC), Singapore PUB, the Institute for Water Quality, Resources and Waste Management, Vienna University of Technology (IWAG-TU) in Austria, Technologiezentrum Wasser (TZW) in Germany, Trojan Technologies in Canada, Water Resources Commission (WRC) in South Africa, and the Water Services Association of Australia (WSAA).

1.3 Project Collaborating Organizations and Roles

This project was sponsored by WERF under the direction of program manager Lola Olabode, M.P.H. This project was funded as part of WERF research on trace organics described previously; however, the project outcome (analysis of risk communication principles and practices) will be broadly applicable to other topic areas. Several other organizations expressed an interest in partnering with WERF to collaboratively sponsor research on this topic, including WEF, NACWA, and the WateReuse Foundation. Linda Kelly of WEF and Cynthia Finley of NACWA participated in project planning meetings and reviewed draft deliverables. Members of the WERF Issue Area Team (IAT) have also provided input and direction on the final scope of work during project meetings and draft document review. IAT members included Amy Woodis of Metropolitan Water Reclamation District, Colorado; Elizabeth Toot-Levy of Northeast Ohio Regional Sewer District; Margaret H. Nellor of Nellor Environmental Associates; and Bill Cairns of Trojan Technologies.

The project was also supported by in-kind contributions in the form of staff labor and other resources from several participating water utilities, wastewater utilities, and water management districts. Staff at these districts, most of who were identified during the proposal phase of work, participated in project team conference calls; shared their experiences and interactions with the media, the public, and regulatory agencies; and shared copies of their communications/communication plans with the project team. They participated in a conference call to answer questions about trace organic compounds and risk communication. They also reviewed and provided comments on portions of the draft report. Utility partners participated from the following organizations:

- Central Contra Costa Sanitary District, California
- City of Boulder, Colorado
- City of Palo Alto, California
- City of Scottsdale, Arizona

- East Bay Municipal Utilities District, California
- Los Angeles County Sanitation Districts, California
- Orange County Water District, California
- Racine Water and Wastewater Utilities, Wisconsin
- South Florida Water Management District, Florida
- West Basin Municipal Water District, California

Project team members included representatives from Malcolm Pirnie, Inc., CH2M HILL, Decision Research, TRG & Associates, and the University of California at Berkeley. Rula A. Deeb, Ph.D., BCEEM and Edward G. Means III of Malcolm Pirnie, Inc. were co-principal investigators. Dr. Deeb was the project manager. Elisabeth L. Hawley, P.E. of Malcolm Pirnie, Inc. served as a task leader. Ms. Linda Macpherson, Vice President and Principal Technologies/Reuse Public Acceptance Specialist at CH2M Hill, was a task leader and project advisor. She was assisted by Jozee Adamson of Decision Research. Other project advisors included Paul Slovic, Ph.D., Professor of Psychology at the University of Oregon and Founder and President of Decision Research; David Sedlak, Ph.D., Professor of Civil and Environmental Engineering at the University of California at Berkeley; Lois Humphreys, President and Owner of TRG & Associates, and Michael C. Kavanaugh, Ph.D., P.E., BCEE, Vice President and Global Science and Technology Leader at Malcolm Pirnie, Inc.

CHAPTER 2.0

METHODOLOGY

2.1 Literature Review

2.1.1 Focused Literature Review

As Task 1, the project team conducted a focused literature review to summarize and synthesize select publications on effective risk communication practices, public perception, and message effectiveness. Most of the documents focused on communication of various environmental risks, including a few focused on trace organic compounds. Several other references were more general, describing risk communication techniques and underlying psychology of risk perception.

Approximately 20 documents were specifically identified by WERF. Several other relevant documents were identified by the project team. Most of the documents were published in the U.S. but a few were from other countries (Australia, Canada). The project team reviewed the following documents:

- WERF, 2003. *Water Reuse: Understanding Public Perception and Participation.* Water Environment Research Foundation, 00PUM1.
- WERF, 2004. *Public Perception of Biosolids Recycling: Developing Public Participation and Earning Trust.* Water Environment Research Foundation, 00PUM5.
- WEF, 2007a. *Microconstituents in the Water Environment*, resource paper for public and press. Water Environment Federation's Microconstituents Community of Practice of the Technical Practice Committee.
- WEF, 2007b. *Microconstituents Glossary.* Prepared by Microconstituents Community of Practice of the Technical Practice Committee.
- WERF, 2008. *Technical Brief: Trace Organic Compounds and Implications for Wastewater Treatment.* Water Environment Research Foundation, CEC3R07. Co-published by International Water Association (IWA) Publishing.
- WateReuse Association, 2004. *Innovative Applications in Water Reuse: Ten Case Studies.*
- WateReuse Foundation, 2006. *Marketing Nonpotable Recycled Water: A Guidebook for Successful Public Outreach & Customer Marketing.* Product No. 03-005-01.
- WateReuse Foundation, 2004a. *Best Practices for Developing Indirect Potable Reuse Projects: Phase 1 Report.* Product No. 01-004-01.
- WateReuse Foundation, 2004b. *Research Needs Assessment Workshop: Human Reactions to Water Reuse.* Prepared by Brent M. Haddad, University of California at Santa Cruz. Co-sponsored by the U.S. Bureau of Reclamation. WateReuse Foundation Report WRF-03-011.
- Perras, J., 2005. *Communications Essentials for Agency Attorneys and Managers.* Perras and Associates, NACWA Developments in Clean Water Law Seminar, November 10, Santa Fe, NM.

- Health Canada, 2006. A Framework for Strategic Risk Communications within the Context of Health Canada and the PHAC's Integrated Risk Management. Chapter 2 in *Strategic Risk Communications Framework for Health Canada and the Public Health Agency of Canada.* ISBN No. 0-662-44596-1. November. Available online at www.riskcommunications.gc.ca
- AwwaRF, 2004. *Risk Communication for Emerging Contaminants.* American Water Works Research Foundation #2776.
- AwwaRF, 2003. *Consumer Perceptions of Tap Water, Bottled Water, and Filtration Devices*. American Water Works Research Foundation #2638.
- AwwaRF, 2005. *Message Management: Effective Communications*. American Water Works Association Research Foundation #2766.
- AwwaRF, 2008b. *State of Knowledge of Endocrine Disruptors and Pharmaceuticals in Drinking Water*. American Water Works Association Research Foundation #3033.
- Roseth, N., 2006. *Community Views on Water Shortages and Conservation*. Research Report No. 28. The Cooperative Research Centre (CRC) for Water Quality and Treatment, Australia.
- Roseth, N., 2008. *Community Views on Recycled Water – The Impact of Information.* Research Report No. 48. The Cooperative Research Centre (CRC) for Water Quality and Treatment, Australia.
- NACWA, 2005. *Pharmaceuticals and Personal Care Products in the Environment: A White Paper on Options for the Wastewater Treatment Community.* National Association of Clean Water Agencies. November.
- Slovic, P. 2007. *Risk Perception and Communication (Lessons from the Risk Assessment Battlefield).* Water Environment Research Foundation Workshop on Trace Organic Compounds: Mapping a Collaborative Research Roadmap, San Francisco, CA, May 17.
- ATSDR, 2008. *A Primer on Health Risk Communication Principles and Practices.* Agency of Toxic Substances and Disease Registry. Available online at www.atsdr.cdc.gov/HEC/primer.html
- Covello, V.T., 2006. *Core slides: Communicating in high concern/low trust situations: Applications to water security*. Center for Risk Communication, NY.
- WHO, 2005. *Effective media communication during public emergencies: A WHO handbook*. Geneva, Switzerland. July.
- AwwaRF, 2008a. *Guidance manual for communicating the value of water*. American Water Works Association Research Foundation #3113.

Elements of effective risk communication practices were extracted from each document and used to develop an organizational framework and the content of this report. Key findings are summarized in Section 3.0.

2.1.2 Literature Review of Other Industries

The project team identified and researched several articles and case studies about risk communication practices and principles in other industries, including the nuclear power industry, health and medical industries, chemical manufacturing, and hazardous waste sites. These topics have aroused strong adverse feelings and reactions in the public and other stakeholders, resulting in careful study of effective communication principles and lessons learned from communication mishaps. This task (Task 2) was guided by the past work and expertise of project advisor Dr. Paul Slovic. The project team reviewed the following documents:

- Cline, D., 1979. *What are Effects? Living under Nuclear Fear*. The Evening Sentinel. April 4.
- Covello, V.T., 1991. *Risk comparisons and risk communication: Issues and problems in comparing health and environmental risks*. In R.E. Kasperson & P.J.M. Stallen (Eds.), Communicating risks to the public: International perspectives. The Netherlands: Kluwer Academic.
- Drew, C.H., Grace, D.A., Silbernagel, S.M., Hemmings, E.S., Smith, A., Griffith, W.C., Takaro, T.K., and Faustman, E.M., 2003. *Nuclear waste transportation: Case studies of identifying stakeholder risk information needs*. Environmental Health Perspectives 111(3): 263-272.
- Flynn, J.H., Slovic, P., and Mertz, C.K., 1993. *The Nevada Initiative: A risk communications fiasco*. Risk analysis 13(5), 497-502.
- Gregory, R., Flynn, J., and Slovic, P., 1995. *Technological stigma*. American Scientist 83, 220-223. Reprinted as Chapter 1 in Flynn, J. Slovic, P., and Kunreuther, H. (Eds.), 2001. "Risk, Media and Stigma," Earthscan Publications Ltd, London, UK.
- Kunreuther, H. and Slovic, P., 2000. *Coping with stigma: Challenges and opportunities*. Reprinted as Chapter 21 in Flynn, J. Slovic, P., and Kunreuther, H. (Eds.), 2001. "Risk, Media and Stigma," Earthscan Publications Ltd, London, UK.
- Leiss, W., 2001. *Dioxins or chemical stigmata*. Chapter 16 in Flynn, J. Slovic, P., and Kunreuther, H. (Eds.), 2001. "Risk, Media and Stigma," Earthscan Publications Ltd, London, UK.
- Long, R., 2001. *An industry perspective on risk and stigma*. Chapter 17 in Flynn, J., Slovic, P., and Kunreuther, H. (Eds.), 2001. "Risk, Media and Stigma," Earthscan Publications Ltd, London, UK.
- NRC, 1996. *Understanding risk: Informing decisions in a democratic society*. Stern, P.C. and Fineburg, H.V. (Eds.), Committee on Risk Characterization, National Research Council, ISBN 0-309-57849-3.
- Purchase, I.F.H. and Slovic, P., 1999. *Quantitative risk assessment breeds fear*. Human and Ecological Risk Assessment, 5(3): 445-453.
- Rozin, P., 2001. *Technological stigma: Some perspectives from the study of contagion*. Chapter 3 in Flynn, J. Slovic, P., and Kunreuther, H. (Eds.), 2001. "Risk, Media and Stigma," Earthscan Publications Ltd, London, UK.
- Sandman, P.M., 2004. *Three Mile Island – Twenty-five years later*. Safety at Work. April 24. Available online at www.psandman.com/articles/3-mile2.pdf.
- Satterfield, T. et al., 2001. *Risk lived, stigma experienced: Reflections on the limits of adaptations*. Chapter 5 in Flynn, J. Slovic, P., and Kunreuther, H. (Eds.), 2001. "Risk, Media and Stigma," Earthscan Publications Ltd, London, UK.
- Slovic, P., 1993. *Perceived risk, trust, and democracy*. Risk Analysis 13(6): 675-682.
- Slovic, P., 1999. *Trust, emotion, sex, politics and science: Surveying the risk-assessment battlefield*. Risk Analysis, 19(4): 689-701.
- Slovic, P., 2007a. *Risk Perception and Communication (Lessons from the Risk Assessment Battlefield*. Presented at the WERF Workshop on Trace Organic Compounds: Mapping a Collaborative Research Roadmap, May 17-18, San Francisco, CA.

- Slovic, P., Peters, E., Grana, J., Berger, S., and Dieck, G.S., 2007. *Risk perception of prescription drugs: Results of a national survey*. Drug Information Journal 41: 81-100.
- TIME magazine, 1979. *A nuclear nightmare*. Monday, April 9. Available online at www.time.com/time/printout/0,8816,920196,00.html
- Van Ryan, A.D., 2003. *Managing trust by bold inclusion*. Environmental Law Newsletter, Fall.
- Walker, 2001. *Defining and identifying "stigma"*. Chapter 22 in Flynn, J. Slovic, P., and Kunreuther, H. (Eds.), 2001. "Risk, Media and Stigma," Earthscan Publications Ltd, London, UK.

Results are presented in Section 3.6. Many of the key findings underscore the same underlying communication principles that were identified during the focused literature review of communications within the water industry.

2.2 Media Analysis and Review

One of the most important ways that public perceptions of health and environmental impacts are shaped is through the media. Newspaper, magazine, television, and movie coverage of the health risks of trace organic contaminants (and similar contaminant-related stories) may convey scientific information, exaggerate, dramatize, misstate scientific information, and/or utilize stigmatizing language that politicize or sensationalize the story.

Under Task 3, the project team identified 26 recently-published media articles and other media coverage that address the issue of microconstituents or trace organic compounds to provide insights into the some of the following questions:

- Does the article convey or express documented scientific risk or create a perception of risk by describing the risk as new?
- Do the media articles use unfamiliar or chemical names that sound threatening?
- Do the media pieces convey information about detection? How is detection described?
- What key sources of information cited in the media pieces? Are additional references cited in the media information?
- Do the media articles contain information about choices and trade-offs?
- What tools or approaches have been used by our key utility partners in the water industry to resolve adverse public reaction or media attention? When were they used and did the utility partner believe there were helpful in attenuating or amplifying concern about microconstituents or trace organics in water?
- Have utility partners identified in this study witnessed any changes in public perception of risks associated with microconstituents over time, as a function of setting (e.g., during a drought when reuse for different purposes, when a drug take-back program is identified)?
- What are the major deficiencies/needs (message content, style, trust, choice of media, etc) identified by our water industry partners that have compromised effective communication by the water industry to the media and to the greater public on these issues? What approaches have been effective? Why?

To answer these questions, the project team used a coding system to characterize the tone of the language and images used (i.e., negative, positive or neutral) and analyzed the results and probable impact of media coverage of trace organic compounds as summarized in Section 4. In addition, the project team interviewed representatives from the 10 utility partners.

2.2.1 Media Article Identification

Several different news articles written from different perspectives or from different locations were reviewed and analyzed. These came from newspapers, journal articles, radio, and television. Media articles were identified by project team members, collected after a series of articles on pharmaceuticals was published by the AP, and/or were provided to the project team by participating utilities and water districts following their interviews. Media reviews were not limited to the U.S. and included articles from Australia, Asia, and Europe, consistent with the task purpose identified above. Articles reviewed included the following:

- Arrandale, Tom, 2008. Flushing Away Fears. Available online at http://governing.com. May 2008.
- Associated Press, 2004. Male fish growing eggs found in Potomac. MSNBC. December 21, 2004. Available online at http://www.msnbc.msn.com/id/6741471/
- Associated Press, 2008. Study finds pharmaceutical drugs in tap water. March 10, 2008.
- Athavaley, Anjali, 2008. Sewer to Spigot: Recycled Water. The Wall Street Journal. May 15, 2008.
- Atteberry, Andra, 2008. Don't flush the pills - drop them off. The Benton County Daily Record. April 10, 2008. Available online at http://www.nwanews.com/bcdr/News/60608/
- Author unknown, 2008. Are you concerned about drinking water? The Atlanta Journal-Constitution. March 26, 2008. Available online at http://ajc.com
- Beeler, Nate, 2008. Editorial Cartoon: Jackpot! The Examiner: San Francisco, p.12. March 11, 2008. Available online at http://examiner.com
- Dean, Cornelia, 2007. Drugs Are in the Water. Does It Matter? The New York Times. April 3, 2007.
- Donn, Jeff, Martha Mendoza, and Justin Pritchard, 2008. Animals Show Effects of Drugs in Waterways - Part 2/3 in an AP Special Investigation. WCBS Newsradio 880. March 11, 2008.
- Donn, Jeff, Martha Mendoza, and Justin Pritchard, 2008. No Regulations to Keep Pharmaceuticals Out of Drinking Water - Part 3/3 in an AP Special Investigation. WCBS Newsradio 880. March 12, 2008.
- Donn, Jeff, Martha Mendoza, and Justin Pritchard, 2008. U.S. Water Not as Pristine as Thought - Part 1/3 in an AP Special Investigation. WCBS Newsradio 880. March 10, 2008. Available online at http://www.wcbs880.com/printyage.php?contentId=1692796&contentType=4
- Eilperin, Juliet, 2006. Pharmaceuticals in Waterways Raise Concern. Washington Post, June 23, 2005. Available online at http://washingtonpost.com
- Friederici, Peter, 2007. Facing the Yuck Factor. High Country News. September 17, 2007.
- Jennings, Simon, 2007. Scientists probe gender-bending chemicals in Potomac fish. Environmental Data Interactive Exchange. January 19, 2007. Available online at http://www.edie.net/news/news_story.asp?id=12494

- Jesperson, Kathy, 2003. Endocrine Disruptors – What are they doing to you? On Tap. Winter 2003.
- McGuire, Kim, 2005. Unused Medicine a "Painful" Waste, Possible Hazard. The Denver Post. April 17, 2005.
- Mendoza , Martha, 2008. Texas town releases name of drug found in water. Associated Press. June 9, 2008.
- Mosier, Jeff. 2008, Arlington looks to eliminate drugs from tap water. The Dallas Morning News. March 26, 2008.
- Mosier, Jeff. 2008, Arlington won't identify pharmaceutical drug found in city's water. The Dallas Morning News. March 10, 2008.
- Pitzer, Gary. 2004, Pharmaceuticals & Personal Care Products: An Rx for Water Quality Problems? Western Water, pp.4-13. July/August 2004.
- Rake, Launce, 2006. Contaminants change fish. Las Vegas Sun. November 4, 2006. Available online at http://trib.com
- Rubenstein, Sarah, 2008. Big Pharma Is in the Water. The Wall Street Journal. March 10, 2008. Available online at http://blogs.wsj.com/health/2008/03/10/big-pharma-is-in-thewater/?mod=WSJBlog/trackback/
- Testicles Shrinking in Las Vegas Bay. Southwest Hydrology, p.31. March/April 2007. Available online at http://pubs.usgs.gov/fs/2006/3131/
- Traces of drugs found in LV Wash. Las Vegas Sun, October 16, 2000.
- Underwood, Anne, 2007. Rivers of Doubt. Newsweek. June 4, 2007.
- Wallar, Michelle, 2006. Flush with chemicals. The Daily Camera. April 17, 2005. Available online at http://www.dailycamera.com/bdc/science/article/0,1713,BDC_2432_3707304,00.html

An analysis of the media coverage is described in Section 4.0 in the final report. Copies of the media articles are provided in Appendix A.

2.2.2 Media Article Coding

A set of coding categories was created based on the literature in risk perception and risk communication. For example, the theoretical model of "social amplification of risk" (Kasperson et al., 2003) centers around the key messages or "signals" emanating from a risk incident or news report of an incident. Thus the project team looked for the signals or themes of each article, to answer questions such as: Is the issue presented as an emerging problem (i.e., as if the risk is new or is increasing)? Does the article indicate that someone is to blame? Are authorities portrayed as competent and trustworthy?

In addition to coding for key messages or signals, the project team looked for other factors that are known to contribute to the social amplification of risk, including a lack of transparency (e.g., does the article present information about risk being suppressed or describe a lack of testing?), limits to the technology being used to identify and construct the risk, scientific understanding of the risk, and the impact of the risk on sensitive populations (e.g., pregnant women, children, or the elderly).

Recent research has identified the importance of imagery and feeling (affect) in guiding risk perception and acceptance (Slovic et al., 2004). Accordingly the project team coded articles to determine what images and inflammatory terms were conveyed in the media articles. The team also judged the overall affective tone of the article (ranging from very negative to very positive). In addition, they coded basic elements of risk, including the pharmaceutical chemicals, named, and whether the name was commonly known (e.g., painkillers) or scientific, the scope of contamination, the amount of exposure to the named chemicals, the conclusions of any risk assessment that were performed, mentions of observable harm, interviews with persons harmed, regulatory actions, and proposed solutions to the problems raised in the article.

A sample article was coded and the coding was critiqued under the direction of task leader Linda Macpherson, project advisor Paul Slovic, and Jozee Adamson, a staff researcher at Decision Research. Ms. Adamson coded each article under the supervision of Dr. Paul Slovic. Whenever possible, the actual text was provided in support of the coded content. For example, in one news report, under the code "scope of contamination" the selected text read: "A vast array of pharmaceuticals have been found in the drinking water supplies of at least 41 million Americans." Additional details about how the coded comments fit into each of the categories will be described in Section 4.0, along with the table of results. The presentation of coded text provides readers the opportunity, if they wish, to consider alternative codes or alternative conclusions to the ones presented in the analysis.

2.2.3 Feedback from Utility Partners

This task involved collecting feedback from key utility partners for information on media impact on perceptions. Several water utilities, wastewater utilities, and water management districts pledged to support this project with in-kind funding. These agencies participated in project team conference calls; shared their experiences interacting with the media, the public and regulatory agencies; and shared copies of their communications strategies with the project team. The following questions were prepared by the project team and approved by WERF and IAT representatives to ask each of the utility representatives:

1. How important of a topic are trace organic compounds within your organization?
2. Does your organization have a designated spokesperson to answer questions from the public about trace organic compounds? Also, does your organization have a designated spokesperson to answer questions from the media about trace organic compounds?
3. Has your organization prepared any outreach materials on the issue of trace organic compounds? If so, please describe these materials and how they were distributed.
4. Have you ever used WEF or WERF's resources on trace organic compounds or microconstituents, such as their website, Technical Practice Updates or Questions and Answers? If so, was the information helpful? What other resources have been useful?
5. What, if any, are the principal concerns in your community related to trace organic compounds (e.g., human health concerns, environmental health concerns, other)? Please estimate the percentage of public concern associated with each.
6. Has your organization developed any partnerships to assist in communicating the risks associated with trace organic compounds (e.g., partnerships with local research organizations, medical doctors, consultants)?
7. What tools, processes, and approaches have your organization used to interact with the public? Describe the situations for which each approach has been used. Do you think that these methods attenuate public concern about trace organic compounds?

8. In your geographic area, have there been any changes in practices or programs to address trace organic compounds, such as monitoring programs, treatment studies or pharmaceutical take-back and disposal programs? Were any of these changes a direct result of public concern? Were any initiated by or supported by your organization?
9. Has your organization witnessed any changes in public involvement or perception regarding trace organics over time as a function of setting (e.g., during a drought, in response to an event, in response to the establishment of a program, etc.)? Please describe.
10. Please identify a local media report (article, radio or television report) as an example of how the media presents information on this issue and how your organization is represented.
11. Has your organization identified any communication deficiencies or needs (e.g., message content, style, trust, choice of media, etc.) that would improve the water industry's communication with the media and with the greater public on trace organic compounds/microconstituents? Similarly, what approaches have been effective? Why?

Utility teaming partners were contacted by telephone for 1.5-hour interviews and were provided with the list of questions in advance of the telephone conversation. The utility representative's responses were recorded and analyzed, as shown in Section 4.0.

2.3 Development of a General Communication Framework

Based on the literature review, a general framework for communications was developed (Section 5.0). Such a framework was prepared to illustrate the steps in the communications process and place actual communications (e.g., brochures, public meetings, tours, webinars) in the perspective of the bigger picture of activities that utilities can undertake to improve the effectiveness of communications. The general communication framework was developed under the direction of Ed Means, by organizing communication principles, practices, and recommendations gleaned from the literature review into several key steps. This framework can be used to convey a wide variety of messages, including those relevant to trace organic compounds as well as a wide variety of other potential health and environmental communication topics.

2.4 Summary and Recommendations

The project team summarized key communication principles and lessons learned from the literature review, media analysis, and utility interviews. The summary was presented in the context of key findings and recommendations for utilities to make use of each finding in the communication process. Next, the project team listed recommendations for further research (Section 6.0).

CHAPTER 3.0

LITERATURE REVIEW

This chapter summarizes the results of a literature review of general risk communication principles, risk communication documents for utilities and summaries of risk communication principles and lessons learned in other industries. These reports described the psychology of risk perception, trust, understanding the public, understanding the media, and a variety of tips for effective communication practices. Sections 3.1 through 3.5 are organized by these topics. Section 3.6 summarizes lessons learned from other industries on risk communication.

3.1 Psychology of Risk Perception

Several articles and reports were reviewed that describe social psychology and risk perception. As described in Section 2.1, the majority of the research summarized in this section was conducted under the direction of leading researchers in the field of risk perception and communication (including Vincent Covello, Peter Sandman, and Paul Slovic), or was referenced in a review by the National Research Council (NRC, 1996).

The task of risk communication often appears daunting to many environmental professionals and utility managers. Programs designed to communicate risk often fail because the expert community focuses on quantifying risk (risk factors, morbidity, and mortality numbers) and explaining the numbers to the public. A growing body of work is examining the factors that explain the public's perceptions of risk and thus influence decision-making and project implementation. Researchers in this area of study are finding a wide range of factors that underlie the public's perception of risk. Technical information can influence the public's response to those factors, but it may not be sufficiently persuasive on its own.

In addition to understanding risk calculations and statistics, a savvy risk communicator today needs to have a basic understanding of the psychology of risk perception in order to better anticipate and understand customers' and other stakeholders' mindsets, decisions, and reactions to messages. This understanding will facilitate two-way communication and ease the frustration that "we" (technical people) often feel when communicating with "them" (a non-technical public), per Fischhoff (1995), who explained the evolving field of risk communication as follows:

- All we have to do is get the numbers right.
- All we have to do is tell them the numbers.
- All we have to do is explain what we mean by the numbers.
- All we have to do is show them that they have accepted similar risks in the past.
- All we have to do is show them that it is a good deal for them.
- All we have to do is treat them nice.
- All we have to do is make them partners.
- All of the above.

Each of these variables is important to understanding perceptions of different health and environmental risks and effectively communicating about risk. Environmental professionals recognize the need to communicate with the public but fear that politics may lead to decisions that are "dictated by public anxiety rather than firmly rooted in science" (NACWA, 2005). Efforts to manage public and media attention on risks to environment and human health can result in a "risk assessment battlefield" (Slovic, 2007).

Risk management issues may bring forth many underlying concerns about societal and community power imbalances (WERF, 2003). Although conflict can be challenging, social conflict is normal, critical, and useful to the society, indicating that society is capable of communicating different values. Discussions can be more productive when stakeholders can interpret each others' emotions and motivations and understand the reasons behind each others' behavior and actions (WERF, 2003). It is therefore advantageous to understand the psychology of risk perception when communicating with others about potential risks to human health and the environment, including the potential risks of trace organic compounds. To facilitate productive and healthy dialogue between stakeholders, several concepts are presented below including risk, risk assessment, "intuitive toxicology," affect, and stigma (Slovic, 2007).

3.1.1 Risk

Among other definitions, risk has been defined as "a threat to that which we value," (Covello, 2007) and as "the probability that something will cause injury or harm" (Agency for Toxic Substances and Disease Registry (ATSDR), 2008). The different focus of these two definitions reflects the different landscapes surrounding the concept of risk to the general public and to risk assessment practitioners. The discrepancy between peoples' perceptions of risk and statistically-measured risk was the focus of much research conducted in the 1980s. These studies identified a list of attributes that influence peoples' perception of risk, including whether or not the risk is voluntary, familiar, whether effects are delayed in time, uncertainty about probabilities or consequences of exposure, severity of outcome, the amount of personal control over the outcome and other factors (Fischhoff et al., 1981; Slovic, 2007).

Despite data illustrating that risk and benefit are typically positively correlated for society (i.e., high risk activities are often undertaken because they have high benefits to society), people's perceptions of risk and benefit are typically negatively correlated (Slovic el al., 2007). When people are asked to rate the risk of an activity, their answer correlates more closely with their perception of benefit or lack of benefit (Slovic et al., 2007). For example, medical exposures to x-rays or pharmaceuticals were perceived to be less risky than nuclear power or pesticides. Researchers' interpretation was that the perceived benefits of medical applications of technologies resulted in a lower perceived risk. The strength of the negative relationship between risk and benefit depended on the degree to which the activity was judged to be good or bad (Slovic et al., 2007).

3.1.2 Risk Assessment

Risk assessment is a step in the risk management process involving the quantification of risk, sometimes expressed as a probability that an adverse outcome will occur. Risk assessment results are used to support policy and other decisions to mitigate or manage risk. Risk assessors (unlike the general public) are familiar with the uncertainty and debates internal to the risk assessment process, including the extrapolation of toxicological data from high doses to low doses, from animals to humans, difficulty with study design, statistical limitations of accuracy, and appropriate safety factors.

In addition to these limitations, factors that are not considered in the formal risk assessment often lead to conflict. Risk assessments do not necessarily address all quantifiable risks. Instead they may focus just on human health and ecological health risks. For example, the impact of contamination on property values is not typically considered or addressed by a risk assessment though it may be quantifiable and important to stakeholders (NRC, 1996). Economic impacts of the Mediterranean fruit fly (MedFly) on agriculture in California in the 1980s were evaluated in a risk assessment (along with the potential human health effects of aerial spraying) (NRC, 1996). However, the risk assessment did not evaluate other risk management alternatives. Therefore, the issue was framed as a choice between the degradation of the agricultural economy and the potential degradation of human health (NRC, 1996). Relevant issues such as the health effects, costs, benefits, likely effectiveness of preventative measures, biological control options, or other methods of spraying were not evaluated (NRC, 1996).

Other issues that are important to the risk management outcome but cannot be quantified during the risk assessment include trust, risk perception, individual experiences and culture, morals, uncertainty, dread, controllability, potential for catastrophe, risk to future generations, power, equity, conflict, controversy, politics, and other factors (Slovic, 2007; NRC, 1996). Risk scholars have learned that effective communication about risky behavior inevitably fails if it is insensitive to the following variables:

- The particulars of how different lay populations perceive and characterize specific risks.
- The way in which different populations use affective judgments and intuitions about toxicological impacts to assess risks.
- The cluster of ideas or images associated with different phenomena; for instance, the mental pictures or word associations that follow mention of the words "chemical," "contaminant" or "emerging concern."
- The way in which different groups within the general public rank and evaluate different risks. (How do different people perceive the human health or environmental risks in water compared with the perceived risks of sunbathing, caffeine, a poor diet, or driving without a seatbelt?)
- The baseline knowledge that different groups already possess about causality or different risk factors associated with mental models.
- The effect on risk perceptions of the level of trust in which the public holds the agency or body responsible for managing a risk.

Because of these factors, some people question the idea that risk (defined as a threat to that which is valued) can be quantified or thought of as a quantifiable reality (NRC, 1996 in review of Douglas and Wildavsky, 1982; Funtowicz and Ravetz, 1992; Krimsky and Golding, 1992; Otway, 1992; Pidgeon et al., 1992; Slovic, 1992; Watson, 1981; Wynne, 1992). The authors point out that the concept of risk is useful to people in navigating uncertainties in life and may be shaped by their individual experiences and culture, i.e., that risk perception and risk tolerance vary with the individual.

This multi-dimensionality of risk helps explain why risk assessments lack authority for some stakeholders and fail to bring them to a consensus. By not addressing these important issues, the information provided by a risk assessment may be seen as invalid, illegitimate or not pertinent (NRC, 1996). Three factors that are important to stakeholders but are often not considered in the risk assessment include fairness, prevention (of pollution or risk), and rights (NRC, 1996). For example, environmental justice issues raise questions about the fairness of exposing a community with low socioeconomic status to an environmental hazard. The community may be less educated/organized/wealthy and therefore less able to effectively voice their concerns or make a case against an environmental hazard. The community may have characteristics that make it unusually susceptible to contamination that contributes to body burden or declining property values. Understanding the psychological definition of risk and incorporating these concepts into the decision-making process will help stakeholders better understand, validate, and respect others' viewpoints and concerns.

3.1.3 Intuitive Toxicology

People tend to rely on a set of intuitions to guide them in determining risk or hazard. The term "intuitive toxicology" was introduced by Kraus, Malmfors, and Slovic (1992) for these beliefs. One of the basic statements describing the science of toxicology is the saying "the dose makes the poison." However, many people lack dose-response sensitivity, intuitively believing that if large exposures are bad, small exposures are also bad (Slovic, 2007). In a survey of students and other groups in the U.S., 26% of respondents agreed that a diet free of salt is more healthful than the same diet with a pinch of salt (Rozin et al., 1996). Similarly, 60 to 75% of people believe that if a person is exposed to a chemical that can cause cancer, that person will probably get cancer some day and that exposure to radiation will probably lead to cancer some day (Slovic, 2007).

Instead of a dose-response model, the public may use a *contagion model* when thinking about exposure to contaminants. In reference to Kai Erikson's work, Slovic described the model as "once in contact, always in contact," i.e., contagion is insensitive to dose (Slovic, 2007). In fact, people may equate very small amounts of exposure with more toxic or hazardous effects. When small amounts of contaminants (including invisible contaminants) are linked to human health effects, these contaminants only seem more insidious and dangerous (Slovic, 2007). The contagion model has helped social scientists explain how toxicology and related jargon increases, rather than alleviates, public concerns and aversion to chemicals (Slovic, 2007).

3.1.4 Affect

Social scientists have also studied the role of affect in risk perception. Affect refers to an intuition, an automatic emotional tone, feeling, or mood associated with a stimulus that a person may or may not be aware of, that assists the person in making rapid judgments and decisions. In his presentation, Slovic used the example of two people passing each other on the street at night and using first impressions and their affective response to gauge whether or not the other person presented any sort of threat (Slovic, 2007). Affect is a key ingredient in rational decision-making, as it conveys meaning to information, tingeing the event with "good" or "bad," and guiding the person's response. Affective response can govern people's decision-making, particularly when they are asked to respond to a situation that is somewhat unfamiliar.

In the context of environmental risks, the role of affect is apparent in people's responses to surveys. For example, as described previously in Section 3.1.1, people will categorize an activity as low-risk or high-risk based on their affective response to that activity (whether they perceive it to be good or bad), a judgment that is strongly influenced by their perception of benefit (Slovic, 2007). Other studies indicate that word choice can affectively influence a person's response, *underscoring the importance of word choice and imagery in shaping initial perceptions of an activity or technology*. This concept is well-recognized and utilized by industry in packaging and advertising; however, the importance of word choice is often overlooked by technical people with scientific backgrounds and under-utilized when communicating messages about environmental risks, including the risks associated with trace organic compounds. For example, the water industry currently uses a number of different terms to describe these compounds, including emerging contaminants, microconstituents, pharmaceutically active compounds, endocrine disrupting compounds, and others. Recently, organizations such as WEF have raised awareness of this issue and the debate within the water industry about the suitability of different terms (WEF, 2007b).

3.1.5 Stigma

Stigma is a negative connotation associated with a sign or marker that people recognize from another context, leading them to regard the stigmatized object or activity as dangerous or taboo (Slovic, 2007). The word implies that people display an excessive negative reaction, due to their past experiences and/or fears. Examples of stigmatized industries/products include the nuclear industry and the recent ban on British beef. Studies have shown that four factors contribute to the development of stigma associated with a risk: if the risk is perceived as 1) uncontrollable, 2) dread, 3) abnormal, and 4) if there are other available substitutes (Slovic, 2007). The presence of trace organic compounds in drinking water supplies may also be viewed by the general public as uncontrollable, dread, abnormal, and unnecessary. If so, the concept of stigma may be useful for understanding people's reactions. A more detailed discussion of stigma is provided in Section 3.6.6.

3.1.6 Implications

Equipped with these concepts from psychological studies, stakeholders can better understand how the public processes information and use this information to develop risk communication guidelines, design focus groups and surveys, interpret the results, and prepare more effective messages and communication tools. This evidence-based approach to risk communication (i.e., making decisions that are supported by evidence from the social and natural sciences and are ideally informed by research specific to the stakeholders and topic at hand) is recommended by organizations such as Health Canada (2006) and WERF (2003). In addition, an evidence-based approach will likely interest scientists and technical staff, who often feel ill-prepared to address risk communication issues with the general public or the media.

However, risk communication with other stakeholders will not be effective unless there is an established foundation of trust. In addition to designing and testing effective messages that are well-informed by psychological research, an organization must build and maintain a trusting relationship with other stakeholders, as described below.

3.2 Foundations for Trust

To trust means to have confidence or faith in, to believe in the honesty and reliability of others, to have certainty based on past experience, and to allow without fear (WordNet, 2008). Trust is therefore a *barometer of past performance and experiences*. Trust is also fragile; surveys have shown that trust-decreasing events have more impact than trust-increasing events and that sources of bad news are viewed as more credible than sources of good news (Slovic, 2007). It is easier to demonstrate risk than an absence of risk. It is therefore easier to destroy trust than to create it, a factor that has been called the "trust asymmetry principle (Slovic, 2007).

The importance of trust in risk communication is highlighted by studies described by Slovic (2007) and is noted in several publications for utility managers (WERF, 2003; AwwaRF, 2008a). These publications describe trust as important to the success of social institutions (AwwaRF, 2008a), as a precursor for business success, and as a facilitator of smooth and successful communication (Slovic, 2007). Slovic notes that without trust, no form of phrasing or presentation is likely to be successful in risk communication.

3.2.1 Corporations Value Trust

Corporations have embraced the notion of presenting themselves as trustworthy, as evidenced by the phenomenon of corporate social responsibility (CSR) marketing. CSR is a term used to describe companies that are committed to operating in a socially responsible or ethical way. The term has allowed companies to brand themselves as ethically responsible (AwwaRF, 2008a). Of the 500 companies listed on the Standard & Poor (S&P) 500 Index, 85% have CSR websites (Asmus, 2006). CSR can refer to many different behaviors and practices ranging from environmentally friendly products and investments to fair working practices for employees. Of these, providing fair employee treatment and protecting the environment are the two most important behaviors for supporting a socially responsible company image (AwwaRF, 2008a). The rise in popularity of CSR attests to the business value of promoting trust among the public, consumers, and clients.

Another effective approach to gain public trust is through corporate honesty, integrity, and responsiveness to public demands and consumer preferences. Responsiveness and integrity are key values that can be incorporated into CSR-type language and policies. This business model may be more realistic in the capitalistic U.S. society than adopting a philosophy of proactively promoting social equity or change. This approach is preferred by some critics of CSR. To quote Peter Sandman: "I am very leery of corporate responsibility, both as a goal and as a message. It suggests that companies ought to decide what the world needs and provide it, whether the world agrees or not. I am much more comfortable with corporate responsiveness, which suggests rather that companies should notice what their stakeholders are asking for and provide it… we need more CEOs [chief executive officers] that follow the rules. But we don't need more CEOs who make the rules!" (Sandman, 2006).

Other critics may see CSR as a marketing ploy or as a way to distract attention away from ethical questions posed by a corporation's core operations. For example, corporate branding platforms focused on benefits to the environment have given rise to a new term "greenwash," meaning a false claim of environmentalism by polluters, giving a positive public image to environmentally unsound practices. In 2005, Green Life published "America's ten worst greenwashers" (GreenLife.org, 2005). Corporations must be careful that sustainability and CSR branding messages are consistent with the organizations' internal behavior and core business principles; else, the company will merely be perceived as hypocritical. Reviewing current business behavior is the recommended first step to building trust, as described in the next section.

3.2.2 Steps to Build Trust

3.2.2.1 Review Current Practices

Organizations must look internally at their own processes, practices, and values to ensure that their behavior is consistent with the key messages that the organization communicates to the public. If not, their messages will not be credible and trust will be undermined rather than strengthened. For example, utilities that are perceived as wasteful of water (or money) cannot successfully communicate messages regarding water conservation (or the need for rate increases) to the public. The messages would be met with distaste and might not be heeded (AwwaRF, 2008a).

Discrepancies between an organization's messages and their behavior are often juxtaposed by the media, as this angle can make a great news story. For example, the president of Coca-Cola was photographed drinking their business rival's product, Pepsi, much to the media's (and Pepsi's!) delight. Taking a good look at internal processes, practices, and values can be thought of as "media-proofing" the organization.

The examination of an organization's key values and current behavior should ideally be performed by a third party to ensure confidentiality and lack of bias. There may be some differences in the review process for public organizations and private entities; however, no distinction between these processes was made in the literature reviewed (AwwaRF, 2008a). Behavior at all levels of the company should be considered, as staff may have a different experience of the organization than upper management. Such a review may assist with the organization's business strategy in other areas and potentially identify ways to reorganize, streamline internal communication and cut costs.

3.2.2.2 Meet Basic Expectations

Another recommended step to build trust is to identify and meet stakeholders' primary expectations (AwwaRF, 2008a). When people's basic expectations are not met, trust will be degraded. A survey of tap water consumers identified the following consumer expectations: 1) tap water will be safe to drink, 2) the delivery system will be reliable, 3) bills will be accurate, and 4) someone will help if a customer has a problem (AwwaRF, 2005). If expectations are met consistently, customer satisfaction and trust will develop as a result (AwwaRF, 2005).

3.2.2.3 Be Honest and Transparent

Transparency is one of five guiding principles of risk management and risk communication identified by Health Canada (2006). The more open and honest an organization is, the more it earns the public's trust, particularly in times of high stress and low trust. An example of the use of transparency to improve trust is described by van Ryan (2003). Johnson & Johnson came under scrutiny for their product after a child died from taking cyanide-laced Tylenol capsules in the early 1980s. The CEO and the board of directors responded by welcoming the media to attend company meetings about the event. By acting as though they had nothing to hide and conducting an immediate investigation of the incident, including all rumors and allegations, Johnson & Johnson quickly re-gained the public's trust and the media moved on.

Transparency (openness and honesty) was one of three main factors found to improve the level of trust and credibility in low-trust, high-stress, relationships (AwwaRF, 2008a). (The other two factors were demonstrating knowledge and concern.) Defying a negative stereotype in any one of these areas is the key to improving perceptions of trust and credibility (AwwaRF, 2008a).

3.2.2.4 Value the Decision-Making Process

Trust can be built through process as well as outcome. People are most likely to trust spokespersons from industry organizations that demonstrate concern and care (AwwaRF, 2008a). One way to demonstrate concern and care is to adopt a decision-making process that includes the public and other stakeholders. An inclusive decision-making process that truly involves other stakeholders can create trust and build collaborative relationships. It is important that participation is meaningful, rather than a charade of being inclusive on the way to a pre-determined solution. In order to be meaningful and more effective at building trust, stakeholder participation ideally begins early in the decision-making process and is continuous (WERF, 2003; WateReuse Foundation, 2004a). Techniques used to force consensus or promote the organizations' preferred outcome will be seen as manipulative and be counter-productive (WERF, 2003). Organizations that seek input from stakeholders in a democratic manner but maintain decision-making authority must manage the public's expectations by clearly explaining their intended decision-making process so stakeholders understand how their input will be used.

By law, community groups are included in the decision-making process at Superfund sites and other cleanup programs that are modeled after the Superfund process. Community representatives may serve as advisory group members with other stakeholders to make key consensus-based decisions about site characterization, conceptual site model development, remedial planning and implementation. Successful case studies touted in news articles and other publications include a site owned by the Alabama Department of Transportation (ALDOT) (van Ryan, 2003) and the Stringfellow site in Riverside County, California. At the ALDOT site, a community outreach group, consisting of nine residents from the impacted community, worked directly with ALDOT and state regulators (van Ryan, 2003). The group was a success because ALDOT and the impacted community maintained an open dialogue throughout the process. The public was encouraged to be a partner in the decision-making process and not viewed as a "necessary unpleasantness" (van Ryan, 2003). At the Stringfellow site, community advocates were eventually brought into the decision-making process, obtained state funding to hire a technical advisor, and greatly contributed to the selection of a final remedy, working in collaboration with the State of California (a potentially responsible party and remediation oversight agency) as well as several industries that had been named as potentially responsible parties (personal communication with Kavanaugh, 2008).

For water and wastewater utilities, the general public and community groups are not the only stakeholders. Others include elected and appointed officials, managers and employees within the organization (AwwaRF, 2008a). Multiple utilities may be stakeholders if there are jurisdictional issues, leading to political and legal questions (WateReuse Foundation, 2006). The interests of all stakeholder groups need to be understood and their input valued during the decision-making process to build trust (WateReuse Foundation, 2004a).

3.2.2.5 Form Collaborations

Trustworthiness can be increased through programs that expand the involvement of reputable third parties and/or expose decision assumptions to testing and verification (NRC, 1996). Collaborations and partnerships with credible individuals and organizations can greatly increase public trust and confidence (ASTDR, 2004; AwwaRF, 2008a). For water and wastewater utilities, appropriate collaborators include health care providers, university researchers, elected officials, non-governmental organizations and other environmental groups. These groups are listed in order of decreasing "trustworthiness" from the perspective of the public surveyed by the study (AwwaRF, 2008a).

Despite the nationwide trend of declining trust in public officials (AwwaRF, 2008a in review of NORC, 2004), utilities can reinforce key messages about water and wastewater conveyed to the general public by partnering with elected officials, assuming that they share the same views (WERF, 2003; AwwaRF, 2008a). Another benefit of such collaboration is more frequent communication between utilities and elected officials, which facilitates financial planning for public utilities and provides officials better access to water quality knowledge, issues, and concerns (AwwaRF, 2008a; WateReuse Foundation, 2006).

Utilities may also benefit from collaborating with each other, assuming that they share the same vision and goals on a particular issue. For example, utilities and other public agencies could collaborate to conduct outreach and education through a shared program, spokesperson or informational materials. Utilities could sponsor expert(s) to conduct research or serve on a blue ribbon panel regarding health issues, water quality, water reuse or other common public concerns (AwwaRF, 2008a; WateReuse Foundation, 2006). Currently, utilities collaborate with each other locally on shared issues and may belong to the same professional organizations. However, there has been little concerted widespread effort among utilities to proactively communicate shared messages to the public (AwwaRF, 2008a).

Despite public trust in institutions, a United Kingdom survey found that people often relied on their impressions of water quality (mainly water turbidity and the content of suspended solids) to decide whether water quality was acceptable (Jeffrey and Jefferson, 2003). It is therefore important to understand public attitudes and opinions so that experts can present information in a way that resonates with people's past experiences and intuitions before designing and circulating risk communication materials, as described in Section 3.3.

3.3 Understanding Public Attitudes

Research on public attitudes is often conducted in order to craft and deliver messages that resonate with public preferences and values. This process can be critical to brand building and may be termed "market research" (AwwaRF, 2008a). This section describes tools that can be used to gauge public opinions and key findings of past studies on water-related topics.

3.3.1 Tools for Understanding Public Attitudes

Commonly used tools for gauging public opinions include surveys, interviews, and focus groups. Surveys are used to collect statistically representative data from one or more groups. In addition to collecting data on a particular topic, most surveys attempt to capture the survey responders' demographic profile, gender, and socioeconomic status, as well as any recent events in the community that may have influenced public opinions. Survey data can be analyzed to determine if there is any relationship between these factors and peoples' responses, and to adjust for confounding factors accordingly. For example, research has shown that people with a lower socioeconomic status are typically less concerned about water quality contaminants (AwwaRF, 2004) and that white males in general perceive activities to be less risky than white females, blacks or Hispanics (Slovic, 1999). A survey may also attempt to measure the effectiveness of different methods of reaching members of the community by identifying community leaders, community organizations, mapping social networks, and asking people what types of media (e.g., internet, newspaper, television) they rely on for information. Online survey tools such as Survey Monkey (www.surveymonkey.com) are used more frequently these days, reducing the time and expense of conducting surveys.

Focus groups and interviews involve spending more time with people to investigate their attitudes on a given topic in detail. Unlike surveys, focus groups and interviews allow interviewers to ask clarifying questions in order to understand people's initial responses. Because this process takes more time than surveys, there are typically fewer participants. Findings are therefore used as case studies or examples, instead of being statistically representative of the general public. Focus groups are typically designed with input from psychologists; responses may be analyzed from a linguistic or cognitive analysis perspective. As with surveys, community demographics may be considered during data analysis.

One approach to data analysis is known as mental modeling. This approach was developed by Morgan et al. (2002). Mental modeling uses in-depth targeted interviews to guide each individual through a series of topics, allowing them to freely express their thoughts and beliefs. People's responses, word choice, and body language are analyzed to determine how they define a problem, what they pay attention to, what they ignore or react to, and what ideals they use to compare new information against (AwwaRF, 2004). The method can be used to assess judgments, shifts in judgments over time, and the reason for those changes (AwwaRF, 2004). The underlying assumption is that each individual has a web of beliefs that they unconsciously use to guide their interpretation of information and decision-making process (AwwaRF, 2004).

AwwaRF conducted a study involving in-depth interviews and used mental modeling to interpret the results (AwwaRF, 2004). The objective of the study was to understand how water utility customers and "experts" (i.e., employees who were familiar with the utility's processes, procedures and communications) view emerging contaminants, gauge their preferences and beliefs about emerging contaminants, and get information about their risk communication efforts (AwwaRF, 2004). Focus groups have also been conducted to gauge people's attitudes towards recycled water and other topics (e.g., see WERF, 2003).

3.3.2 Key Findings

A number of studies published in the literature on risk communication and perception of environmental issues focus on understanding public opinions and perceptions. Topics include broad environmental issues, water conservation and shortages, tap water, recycled water, and biosolids, as described in the following sections.

3.3.2.1 Environmental Issues

A nationwide Harris (2005) poll of 1,217 adults found that environmental issues are still important to the general U.S. public; 74% agreed that protecting the environment is so important that requirements and standards could not be too high and improvements should be continued, regardless of the cost. The public's top two environmental concerns are the pollution of rivers, lakes, and reservoirs and the pollution of drinking water. In the survey, 71% of people polled believed that they didn't have to choose between economic growth and environmental quality (AwwaRF, 2008a in review of Harris, 2005). This finding underscores the public's expectations for clean water and support for environmental action.

3.3.2.2 Water Conservation and Shortages

CRC Australia recently conducted a survey of people's attitudes towards drought, water conservation, and other related topics (Roseth, 2006). They found that the majority of people polled viewed water as a scarce resource that should be conserved at all times, not just during periods of drought or hot weather (Roseth, 2006). The majority did not believe that it was their right to use water whenever they wanted or that paying for water entitled them to use as much as they wanted. Community members understood that a number of complex factors led to water shortages, including population growth, an absence of water recycling, and an absence of conservation measures. Approximately one-third of the survey participants trusted the authorities to ensure future water supply and one-third were worried about water shortages. However, in-depth interviews of this group revealed no sense of urgency or strong worry. The authors therefore concluded that concern was not strong enough to galvanize greater community action than is already occurring (Roseth, 2006).

In general, women were more concerned about the situation than men, and people with higher education levels were also more concerned. People were motivated to conserve water in order to avoid running out of water, care for the environment and, (on a smaller scale) save money. Most (88%) believed that individuals could make a difference in the amount of water saved and that they personally (79%) could do more to save water. (22% thought they could do a lot more.) Most Australians (70%) considered a garden to be important for quality of life, and agree that a garden must be green to be healthy (Roseth, 2006). Approximately 16% of survey respondents said that having to save water detracted from the quality of their life. Educational materials may be helpful, as 20% said they did not have enough knowledge about what to do to save water (Roseth, 2006).

Attitudes about water may be somewhat different in Australia compared with some parts of the U.S. where supply issues have been less extensive. However, attitudes may be similar in arid regions in the U.S. and areas facing water scarcity. For example, people who live in Arizona or Nevada may recognize that they live in a desert and be more willing to landscape with native shrubs instead of planting and maintaining a lawn; in other regions, such as the southeastern

U.S., lawns may be thought of as necessary. Studies in the U.S. indicate that the public is aware of water supply problems and is interested in having a meaningful involvement in finding a solution (WERF, 2003).

3.3.2.3 Tap Water

A study published by AwwaRF, *Consumer Perceptions of Tap Water, Bottled Water, and Filtration Devices*, investigated public perceptions of tap water in the U.S. (AwwaRF, 2003). This study was motivated by concern over tap water (86% of Americans are concerned about their tap water) and the growing trend of drinking bottled or filtered water (40-50% of Americans drink bottled or filtered water) (AwwaRF, 2003).

By conducting a telephone survey of bottled and filtered water users, researchers found that safety/healthiness was the primary reason for not drinking tap water; this concern was stronger among filtered water drinkers than bottled water drinkers. Taste was a stronger factor for bottled water drinkers, with offensive tastes reported including musty/earthy, metallic, and chlorinated flavors (AwwaRF, 2003). Other studies have shown that 50% of consumers have a negative opinion of chlorine taste and odor in drinking water and the associated healthiness (AwwaRF, 2008a). This is consistent with the finding that people tend to tend to trust their own senses and judge value based on taste, color, and odor (AwwaRF, 2004). Younger consumers (29 and under) were less satisfied with tap water safety, healthiness, and overall quality; satisfaction increased with age (AwwaRF, 2003). Safety of tap water is more important in households with children under age 10 and taste is more important to older adults (AwwaRF, 2005).

3.3.2.4 Recycled Water

Several studies have been conducted to examine the public's views on water recycling (e.g., Roseth, 2006; WERF, 2003). An urban Australian case study was conducted using multiple methods to explore community attitudes towards the use of recycled water (Roseth, 2006). In Mawson Lakes, South Australia, non-potable reuse is practiced for garden watering, toilet flushing, and car washing, through a dual water supply system. Three surveys of the community were undertaken: two prior to the start of the reuse project and the third after the project had begun. Communication, trust in the water authority, water quality, value, and risk perception were found to be important for promoting satisfaction with recycled water (Roseth, 2006).

The WERF study found that people in the U.S. support the general idea of water recycling because water reuse promotes conservation, provides environmental protection and distributes a valuable and limited resource. However, people's support for recycling projects in their community is typically lower than their support of general water reuse projects (WERF, 2003). People are also more willing to reuse their own wastewater (i.e., at a household scale versus a community scale) (WERF, 2003). Public acceptance of water reuse is high when the following characteristics apply:

- The degree of human contact with recycled water is minimal.
- The protection of public health is clear.
- The benefits of reuse are clear (e.g., protection of the environment, promotion of water conservation).
- The cost of treatment and distribution is reasonable.
- The perception of wastewater as the source of reclaimed water is minimal.

- The community has a high awareness of water supply programs.
- The role of reclaimed water in the overall water supply is clear.

Finally, the perception of the quality of reclaimed water is positively correlated with confidence in local management of public utilities and technologies (WERF, 2003). Concerns were noted to be less for non-potable reuse projects versus potable reuse (WERF, 2003).

3.3.2.5 Biosolids

Significant findings from a biosolids knowledge and perception survey include the following (WERF, 2004):

- Most people think that biosolids recycling presents a relatively small risk. People favor constructive uses of biosolids (creating energy, recycling nutrients into the soil, and recycling a necessary waste).
- The choice of words makes a significant difference: using the term "sewage sludge" instead of "biosolids" creates a marked drop in positive response to a neighbor's hypothetical use of the material. Knowledge of the term "biosolids" is limited.
- Some factors increase people's concerns about biosolids (e.g., if biosolids are coming from an industrial waste source or large city) and others decrease concerns (e.g., annual certification process, prior knowledge of biosolids use, local supervision of use).
- Regional differences are minimal.

In general, the survey found that people have questions about biosolids recycling and want more information about many different aspects of it. Although they clearly expressed support for the concept of recycling biosolids, respondents indicated that the strongest argument against biosolids recycling is "not enough is known," indicating that they are uncomfortable with their own lack of information and/or what they perceive to be an overall lack of information on this topic (WERF, 2004). Faced with a hypothetical situation in which biosolids are used by a neighbor, people say they would turn to and trust friends and neighbors, government agencies, and academic researchers for initial information. They would trust those who appear knowledgeable and objective and would strongly distrust those with a profit motive (WERF, 2004).

3.4 Understanding the Media

Information that people receive through the media can strongly influence their ideas about contamination, water resources, local government, and other topics of interest to utilities. Media coverage does not create public opinion, but it can amplify and direct it. It can be a powerful force in framing key issues and play a critical role in how utilities are perceived. Data analysis indicates that the frequency of news media coverage of a water quality topic and whether or not a population of vulnerable people is involved are the two most important attributes in predicting whether a utility conducted risk communication on that particular issue (AwwaRF, 2004). Utilities and other interested parties would benefit from learning how to better communicate with the public through the media. To do this requires an understanding of media interests and motivations, format restrictions, information sources, and the impacts of media coverage, including positive press.

The U.S. news media is fundamentally motivated by financial interests. News corporations are run as businesses and are interested in attracting a larger audience to increase advertising revenue, which is their primary source of income. Because of intense business pressures in the marketplace, more media corporations are interested in "hard" news (accidents, fires, crime scenes, etc.), resulting in media sensationalism and the use of wire copy reports distributed by larger organizations (e.g., the recent Associated Press series on pharmaceuticals in drinking water supplies), increasing the impact of a single news story that is widely distributed (Brown, 2004). As their organizations attempt to keep costs low, local journalists may also not have the time to truly understand the issues before preparing the story and may not have prior experience as a science writer.

Due to the need to hook and retain viewers, news stories on water quality will not be given detailed coverage on television. For television news, these stories are typically sixth or seventh in the news lineup, meaning they are less than one minute long (Perras, 2005). The news media is not interested in educating the public but in collecting a variety of facts and opinions to use in preparing a story. Information about water is often ignored by the media unless it reaches a crisis situation. Many stories involve controversy and give mostly one point of view. The media is typically interested in human interest stories, people's perspectives, and simplified, often polarized, answers (ATSDR, 2004). More heterogeneous communities are likely to have news coverage that frame drinking water issues as problems without mention of solutions (AwwaRF, 2004). Newspaper stories may be longer and more balanced; however, more people get their information from television (Perras, 2005). In addition, people rate television news sources as the most believable; the top five most believable news outlets include 60 Minutes, CNN, NBC News, ABC News, and local TV news (AwwaRF, 2008a). Although newspaper circulation is falling and internet use is rising, local and national newspapers were rated as more believable than other internet news sources (AwwaRF, 2008a).

Understanding the makeup of a story and what interests the media also helps organizations prepare for and respond to questions from the press (AwwaRF, 2004). Today, communication with the media is part of a utility's job (Perras, 2005). The best way to prepare for television coverage is to develop simplified messages that can be delivered in seven seconds or less (Perras, 2005). When fielding and answering questions, Perras recommends keeping the following tips in mind:

- Stay focused on the key messages. (Stick to what's important.)
- Listen to what the public says.
- Make it easy for the public to participate.
- Develop clear written materials.
- Be prepared.

More details on effective communication practices are provided in Section 3.5.

Utilities can refer members of the public and the press to the utility's website for more detailed information. Journalists often rely on websites to identify appropriate contact persons and get other information when writing a story. To encourage journalists to contact a particular spokesperson or department of a utility for information, some utilities have a designated page on the website for journalists (Perras, 2005). From the public's perspective, customers indicated that utility websites were most helpful for getting drinking water safety information, emergency instructions, and emergency contact information (AwwaRF, 2008a). Customers were least interested in website content regarding how the utility was managed and how water was processed and delivered (AwwaRF, 2008a). However, a website is one of the least effective ways to reach customers with news; more effective methods of alerting customers to new information includes printing information on the water bill, including an insert with the water bill (not as widely read as information printed on the bill itself) and being featured in television stories or advertisements (AwwaRF, 2005). Through their website and other resources, utilities can develop positive relationships with local media, which can lead to positive press, news articles, and educational information for the public (AwwaRF, 2008a). More details on media coverage and interaction are presented in Section 4.0.

3.5 Effective Communication Practices

Effective communication practices include developing a communication vision, preparing a communication plan, and implementing that plan effectively. A variety of tips and guidance for effective communication have been published for utilities. Key references include *Communicating the Value of Water* (AwwaRF, 2008a) and *Marketing Nonpotable Recycled Water: A Guidebook for Successful Public Outreach & Customer Marketing* (WateReuse Foundation, 2006). References on effective communication, written for a general audience, include *A Primer on Health Risk Communication Principles and Practices* (ATSDR, 2004) and *Effective Media Communication during Public Emergencies: A World Health Organization (WHO) Handbook* (WHO, 2007), which describes guidelines for communicating in high-concern, low-trust situations, many of which are practices that are generally advisable. The following sections summarize key points from these documents as well as others.

3.5.1 Developing a Communication Vision

The first step in effective communication is to develop a communication vision. This entails characterizing the situation (AwwaRF, 2004) (also referred to by Health Canada (2006) as "defining the opportunity"), identifying dialogue goals, and developing key messages. Literature reviewed did not include any published examples of communication visions, suggesting that utilities: a) may not be placing proper importance on this process as a step in communication planning, b) may be going through this process without documenting it, or c) may be conducting this analysis for internal use only. Defining a communication vision is similar to a corporation's process of developing a brand. Brands are, in a sense, a type of communication vision. Brand development and consulting is a well-established area of market research. Brands are used to communicate key messages to customers consistently in an easily recognizable way. They can help with product identification, signal quality levels, and add value or social status to a product. When developing a brand or other type of communication vision, the goal is to identify and develop a few key messages that are reinforced in all future communications (AwwaRF, 2008a).

Companies with strong brands can be easily identified by just a few words. For example, after hearing the words *individual, performance,* and *irreverence*, many people would correctly identify this as a description of the Nike brand (Goldman and Papson, 1998). Apple is another example of a company that is well-branded and easily recognizable. In the case of Apple, the brand represents imagination, innovation, and design from the point of view of "righteous outsiderism and rebellion against injustice" (Kahney, 2002). According to Naomi Klein, author of *No Logo,* "people are drawn to these brands because [the brands] are selling their own ideas back to them" (Kahney, 2002). People can develop a collective emotional bond with a brand and become loyal supporters. The success of organizations such as Apple and Nike illustrate the power of simple branding, the influence that brands have on price and willingness to pay, and impact on corporate, product, and organizational success.

Successful brands are consistent, excel at delivering what the customer really wants, remain relevant over time, and result in brand loyalty (AwwaRF, 2008a in review of Keller, 2000). Successfully branded companies measure any improvement in performance or product from customers' perspective and price their products and services in a way that is consistent with the consumer's perception of value. Such brands must be sustained over time by proper support and a full repertoire of marketing activities. Brands have equity, which is measured by the degree of brand name awareness, brand loyalty, perceived quality, brand associations, and other assets (patents, trademarks). Successfully branded companies monitor their brand equity over time (AwwaRF, 2008a in review of Keller, 2000).

When developing key messages for their communication vision, utilities must consider their own goals (i.e., what they are trying to convey to their audience), as well as what resonates with their audience. One of the most frustrating experiences for utilities when developing messages about trace organic compounds is the lack of complete understanding about human health impacts, ecological effects, sources, treatability, and other issues, and the need to communicate in the face of uncertainty. Utilities may feel that they do not have enough understanding to formulate key messages. Or, it may seem wrong or immoral to develop "branded messages" instead of just selling the public good-quality water the old-fashioned way. Still, it is important to understand that, in the absence of planned communications, messages are still being communicated to the public on this issue from other information sources, such as the news media. The public will likely get incomplete information and may develop a negative image of water utilities as a result, also termed "negative branding." Brand and image are linked; perception is reality.

Some examples of negative impressions that people can develop or have developed about utilities include the following:

- Utilities are not our protectors.
- Utilities do not have communication ethics.
- Money and politics are more important to utilities than public health.
- Utilities manage water quality recklessly, with "band-aid" approaches to treatment.
- Us vs. them.
- Water systems are outdated.
- Regulations are not adequate.
- People should be drinking bottled water (AwwaRF, 2008a).

Communication visions can be used to develop messages that replace or counteract some of these negative impressions. The vision needs to be simple, memorable, and consistently communicated. Messages must be targeted and the entire process must be dynamic, with market research defining and refining the messaging campaign (AwwaRF, 2008a). For water utilities, the objective of the communication vision is to convert customers into supporters (AwwaRF, 2008a). Traditionally, water and wastewater utilities have tended to stay in the background in marketing and public communication. Professional organizations or trade associations may provide better examples of communication visions for utilities to follow. For example, the California Avocado Commission (CAC) recently developed a marketing campaign on avocados that is well-branded. CAC images and vocabulary consistently conjure up ideas of freshness, nutrition, health, and the California lifestyle (AwwaRF, 2008a; www.avocado.org). Their image is supported by continual marketing and professional outreach (AwwaRF, 2008a).

3.5.2 Preparing a Communication Plan

There are several reasons for utilities to prepare a communications plan. First, the plan reflects the organization's commitment to the importance of risk communication. Preparing a plan will likely include establishing a budget for staff or consultants to develop and implement public outreach and marketing activities (WateReuse Foundation, 2006). Second, the communication plan thought process is an important exercise for risk communicators and other stakeholders to get them thinking about the importance of communication, trust, and other topics. Third, the communication plan helps the organization communicate its values to staff and incorporate staff goals, values, and activities back into the plan. A communication plan can be effective in changing attitudes and practices within the organization (as described by Long, 2001 in a review of Dow Chemical's practices) (Section 3.6.1).

Several reports have published diagrams, lists of communication principles and frameworks for preparing a risk communication and outreach program. An AwwaRF report suggests that a communication plan include responses to four questions:

1. Who are we?
2. Where are we now?
3. Where do we want to go?
4. How will we get there? (AwwaRF, 2004).

The International Association for Public Participation (IAP2) for public involvement provides self-assessment worksheets for utilities to complete as a starting point and summarizes a number of public outreach principles (WERF, 2003):

- Manage information for all.
- Maintain individual motivation and demonstrate organizational commitment.
- Promote communication and public dialogue.
- Ensure fair and sound decision-making.
- Build and maintain trust.

Success in applying these five principles is monitored and changes in outreach and the communication plan are made accordingly. Health Canada, in their document on strategic risk communication (defined as "a purposeful process of skillful interaction with stakeholders supported by appropriate information"), also presents five guiding principles for developing a communication plan (Health Canada, 2006):

1. Strategic risk communications are integral to risk management.
2. Stakeholders are the focal point.
3. Decisions are evidence-based, tapping both social and natural sciences.
4. Risk management and risk communications processes are transparent.
5. There is continuous improvement through evaluation.

Regardless of how the communication plan is structured and how the goals are defined, the common focus of most publications includes 1) looking internally at the organization's commitment and current practices; 2) identifying the stakeholders and understanding their values, interests and priorities; 3) defining the issue, goals, and vision of the communication program; 4) preparing a communication plan; 5) developing and pre-testing messages; 6) implementing the plan while engaging stakeholders; and 7) monitoring the plan's effectiveness. These steps are summarized in Table 3-1, along with the guidance language from each reference. Despite this common framework, the content of a communications plan ought to vary from one organization to another. Following a recipe approach to developing a communications plan would discourage creative thinking and may not accurately reflect stakeholders' interests and priorities. Some examples of communication plan elements include the following:

- Provide internal training (WERF, 2004; WateReuse Foundation, 2006).
- Base communication plans on information, not assumptions. Conduct market research if the budget allows (AwwaRF, 2004; WateReuse Foundation, 2006; see Section 3.3 for further discussion).
- Integrate stakeholders into the planning process (ATSDR, 2004; WateReuse Foundation, 2006; WateReuse Foundation, 2004a; WERF, 2003).
- Include a timeline that is updated annually with plans for outreach to each stakeholder group (WateReuse Foundation, 2006).
- Clearly articulate the problem (ATSDR, 2004). Develop a strategic plan that includes "an analysis of the need for [the project]; budget, public outreach, and marketing goals; challenges and opportunities; key outreach messages; identification of stakeholder groups (including the general public, elected officials, government agencies, media, internal staff, recycled water customers); strategies for stakeholder groups; and recommended communication and advertising materials" (WateReuse Foundation, 2006).
- For water reuse projects, evaluate alternatives to indirect potable reuse with the public (ATSDR, 2004) and have a meaningful supply planning criterion (WateReuse Foundation, 2004a).
- Develop an outreach and marketing plan (WateReuse Foundation, 2006).
- Develop effective messaging, enhance customer awareness of the consistency and reliability of service, and increase overall knowledge and understanding of water quality issues (AwwaRF, 2004; AwwaRF, 2008a; ATSDR, 2004).
- Understand, anticipate and address environmental justice issues (WateReuse Foundation, 2004a).
- Use a variety of channels to deliver the message to the public (AwwaRF, 2005; WateReuse Foundation, 2006).
- Hire effective spokespersons (ATSDR, 2004).

- Address under what circumstances communication with the public or media is warranted (AwwaRF, 2004).
- Communicate with the media (WateReuse Foundation, 2006; ATSDR, 2004; AwwaRF, 2008a; WateReuse Foundation, 2004a).

Table 3-1. Summary of Communication Plan Components.

Component	International Association for Public Participation (IAP2) guidelines (WERF, 2003)	Strategic Risk Communication Framework (Health Canada, 2006)	Risk Communication for Emerging Contaminants (AwwaRF, 2004)	Communicating the Value of Water (AwwaRF, 2008a)	Public Perception of Biosolids Recycling (WERF, 2004)
Internal			Draft the project goal, form the team.	Establish the current utility behavior.	Build commitment within your organization
Public and other stakeholders	Identify what the public values.		Define interests and priorities of key stakeholders.	Identify what the public values.	Determine who the stakeholders are.
Define issue, goals, and vision		Define the opportunity. Identify the issue and its context. Focus current understanding.	Characterize the situation. Define the outreach and dialogue goal.	Establish the desired utility brand and communication vision. Develop communication goals.	Understand the national and local context.
Plan	Determine how a product or item addresses those values. Craft a communication strategy that emphasizes those values and the products' connection to those values.	Develop communications strategies that are sensitive to stakeholders' current thinking, goals and choices. Select a strategy.	Develop the plan, messages, and materials.	Establish behavioral needs and gaps to achieve goals. Evaluate the existing situation and future scenarios. Identify needs and gaps to achieve the communication goal. Prepare communication plan and integrate into strategic plan.	Work on factors you can control. Build "public relationships" with individual stakeholders. Pay attention to trust, fairness, honest and useful information, two-way communication, and organizational commitment. Get input from a variety of stakeholders early in the process, via public participation.
Pre-test messages		Develop and pre-test strategies, risk communications plans and messages.			
Engage stakeholders and implement plan		Implement risk communications plan; assess stakeholder perceptions of the risks, benefits, and tradeoffs; assess how stakeholders perceive the options.	Engage through dialogue.	Implement and monitor communication and strategic plans.	Provide information that is useful to recipients. Improve communications, especially listening and dialogue. Redefine success. Don't be afraid of conflict – work patiently through it.
Monitor effectiveness		Evaluate the risk communication process and effectiveness.	Measure effectiveness.		Continue to monitor public perceptions and relationships.

- Pay attention to the implementation of a new project to avoid snags that could affect customers (WateReuse Foundation, 2006).
- Position a water utility to assess emergency or crisis communication issues (AwwaRF, 2004; Covello, 2006). Be honest if mistakes occur and have a crisis plan in place in advance so everyone knows what to do if there's a problem (WateReuse Foundation, 2006).
- Evaluate communication efforts (ATSDR, 2004).
- Keep the risk communication strategies and plans dynamic; they will need to be updated and adjusted to fit societal changes (AwwaRF, 2004).

More details are provided in the original references and several of these topics are described in more detail in Section 3.5.3.

A conceptual model for thinking about communication elements was prepared by a working group at the 2007 WERF *Workshop on Trace Organic Compounds: Mapping a Collaborative Research Roadmap*. Participants prepared this conceptual model as a way to organize and clarify the discussion of research needs on communication-related topics and provide people with a common framework for thinking about the variety of steps involved in successful communication.

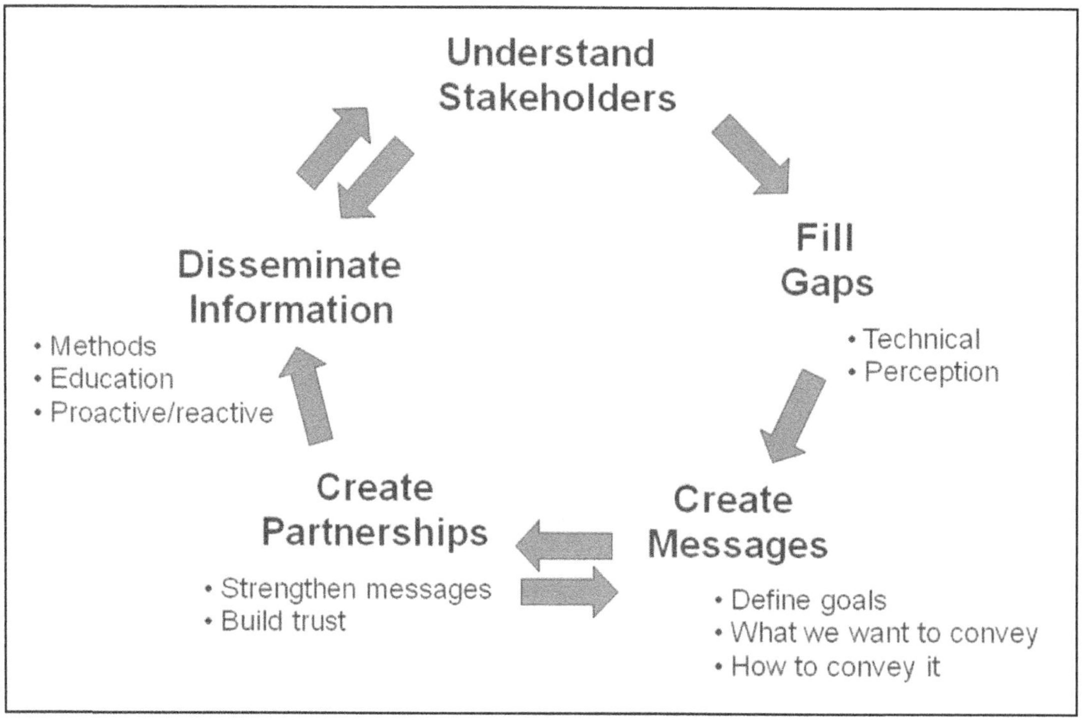

Figure 3-1. Conceptual Communication Model (WERF, 2007).

This schematic illustrates the iterative and adaptive nature of successful communication, as information from each step in the framework feeds into the next step of the process. There are several steps in the model, including: 1) understand stakeholders, 2) fill gaps, 3) create messages, 4) create partnerships, and 5) disseminate information. Key activities undertaken during each of these steps are described in more detail in Section 5.0, which describes a similar, revised, general communication framework.

3.5.3 Implementing the Communication Plan Effectively

The smooth implementation of communication plan elements is critical to the success of a project. Details on several of the key components of the communication plan are provided in this section. Topics include internal training, stakeholder participation, messaging, message delivery to the public, effective spokespersons, communication with the media, smooth implementation (water reuse projects), and guidelines for high-trust/low-concern situations.

3.5.3.1 Internal Training

Several reports highlighted the importance of internal training about risk communication, particularly with respect to biosolids management (WERF, 2004) and water reuse (WateReuse Foundation, 2006). Typically, water and wastewater engineers and managers are not trained in public interaction. However, technical people are often in a position of interacting with the public as part of their daily job (during field work, at community events, answering phone calls) and respond to the public's questions about these issues. Training staff on basic communication principles and preparing them to handle frequently asked questions from the public would be a valuable activity for utilities. Training staff is key to developing stronger public input and participation in biosolids management (WERF, 2004). Training has the added benefit of facilitating communication between different departments within an organization (e.g., marketing and engineering, or water and wastewater) and getting feedback from staff internally, who are often the "eyes and ears" of the program in the community (WateReuse Foundation, 2006).

3.5.3.2 Stakeholder Participation

Utilities are frequently urged to integrate stakeholders into the planning and decision-making processes. This serves to engage citizens, customers, and others early on; listen and respond to their concerns; and accept them as a partner in the planning process (ATSDR, 2004; WateReuse Foundation, 2006; WateReuse Foundation, 2004a). Some people define success in terms of outcome; however, success in communications is typically a combination of outcome, process, and relationship (WERF, 2003). One recommendation is to ask stakeholders to each define success of a project and share their definitions with others (WERF, 2003). Opening a dialogue with stakeholders early in the process builds trust. Frequent communication with regulators is also urged so that there are no unexpected hurdles or obstacles along the way (WateReuse Foundation, 2006).

3.5.3.3 Messaging

When developing messages, consider how the public will be affected by the knowledge/information presented, discussion of benefits or added value, and the choice of vocabulary and imagery.

When communicating with the public, a general rule of thumb is to inform, not educate (WateReuse Foundation, 2004a). Keeping this purpose in mind will help the writer to connect with the public and will provide them with the knowledge they need to make an informed decision without sounding patronizing or confusing or alienating them with too many details. Information provided in context is easier to understand. Costs, concentrations, and other numbers are best presented in meaningful terms that give people a sense of scale or comparison (WateReuse Foundation, 2004a). For example, the estrogenic activity of wastewater secondary effluent was described as having significantly less than (1/1000th of) an equal amount of cow's milk or soy milk (WERF, 2008). Similarly, when describing cost, utilities could discuss rate increases within the context of water rates for peer cities or service areas or present water quality data in the context of an ongoing water quality plan (e.g., watershed protection programs, source control initiatives, treatment programs, and research studies). Monitoring data or treatment system performance could be presented in the context of the utilities' track record (WateReuse Foundation, 2004a).

Research studies have tried to quantify the impact of giving people additional information. For example, one study found that women, nonwhites, urban, and less affluent people have a higher need for information compared with their counterparts (AwwaRF, 2004). The credibility of the source of information was also quite important to these groups (AwwaRF, 2004). Depending on the community demographics and interests, utilities might want to make technical details available to the public to back up information that they distribute (AwwaRF, 2008a). Finally, keep in mind that information is not typically helpful in swaying public opinion but is provided so that people have enough information to come to an informed opinion. For example, one study showed that providing people with additional information does not necessarily make them more receptive to water reuse projects (WateReuse Foundation, 2004a).

People are more supportive of an activity, service or program when the value and benefits are made clear (ATSDR, 2004; WateReuse Foundation, 2004a). People are also more willing to accept risks when they understand the benefits, as discussed in Sections 3.1.1 and 3.1.4. It is a good idea for utilities to emphasize value and benefits in their messaging. For example, when describing recycled water options, the California utility East Bay Municipal Water District (EBMUD) explained to customers that if things continued without any additional water supply, customers would have to cut back on their water usage by 65% in times of drought. With the additional water supply, customers would have to cut back by no more than 25% (AwwaRF, 2008a). Calculations like this create a perception of improvement, as recommended by ATSDR (2004).

Value can sometimes also be expressed in terms of cost. Competitively pricing recycled water is often a major reason for attracting customers; the price of water can be a premier marketing tool (WateReuse Foundation, 2006). The value of tap water compared with bottled water can be illustrated with statistics describing the number of glasses of tap water that cost the same as a single bottle of water. The American Water Works Association has published a compendium of example comparisons and statistics that may be useful for utilities in designing their communications to demonstrate value (AwwaRF, 2008a). In general, upbeat and positive messages about quality, reliability, price, conservation, and source water protection are favored by the public (AwwaRF, 2008a).

The recent AwwaRF project on communicating value of water identified five different categories of value: intrinsic, societal, environmental, economic, and value to users (AwwaRF, 2008a in review of Roger et al., 1998). In addition, the value of water is situational. It could encompass the services associated with uninterrupted delivery of water as well as the value of water itself. The value of the delivery service is typically taken for granted by customers;

however, this value becomes more apparent when there are problems with the distribution system. Currently, in the U.S., people typically undervalue water and therefore are less motivated to use it wisely or pay more for it. High quality, low cost water is taken for granted, rather than viewed as a privilege (AwwaRF, 2008a). This may make it difficult for utilities to get support for capital projects to improve water quality (AwwaRF, 2008a).

The vocabulary and imagery used in communication materials can also have a profound effect on the public's understanding and ideas about the situation. In a recent study, the term "emerging contaminants" was linked in at least one person's mind with the word "emergency", and thought to imply water supply threats related to terrorism (AwwaRF, 2004). This inevitably raised the question of whether or not the water was safe to drink without getting sick (AwwaRF, 2004). Although not statistically significant, this is presented as an example of the types of associations that people can form when different vocabulary is used to describe the topic. Several groups, including the WEF COP on Microconstituents, have extensively discussed different terms for describing these compounds, including contaminants of potential or emerging concern, microconstituents, micropollutants, trace organic compounds, pharmaceuticals and personal care products, pharmaceutically active compounds, endocrine-disrupting compounds, and several other terms. However, the impact of these terms has not yet been tested using surveys or other market research tools.

WEF has developed TPUs or fact sheets which emphasize the importance of vocabulary in conveying or obscuring meaning as well as positive, negative or neutral implications (One descriptive term that was found to be easily understood by the public was "potential pollutants" (AwwaRF, 2004).) WEF has also developed a glossary of terms and definitions (WEF, 2007b). The WateReuse Foundation recently released a request for proposals (RFP) with the ultimate goal of developing an appropriate common vocabulary to create the greatest degree of understanding about water reuse and desalination. This project will identify words and images to do this and will presumably tackle the topic related to trace organic compounds, microconstituents or pharmaceuticals and personal care products in water as this topic impacts water reuse and desalination. The project will be co-funded by the Association of Metropolitan Water Agencies (AMWA), NACWA, WEF, AWWA, U.S. EPA, and the WateReuse Association. The WateReuse Foundation is in the process of selecting a research contractor (WateReuse Foundation, 2008). More resources are expected to be available for utilities on this topic in the future.

3.5.3.4 Message Delivery to the Public

An organization will likely use multiple channels and methods to deliver their messages to the public, in order to reach a wide variety of people. Age, socioeconomic status, and other factors may be correlated with people's preferred source(s) of information (as well as what they value). For example, AwwaRF (2005) found that households with seniors were more likely to read information that was mailed to them along with their water bill. However, the impact of water bill inserts on overall understanding of key issues has not been confirmed. Seniors were also more likely to value having a personal relationship with a utility representative and to gauge the value of water based on taste (AwwaRF, 2005). In contrast, young people often absorb messages at schools. Incorporating water or wastewater education into local schools would be an effective way to reach the younger generation (WateReuse Foundation, 2006). Low-income households value information about the taste of their water and fees; high-income households are mostly interested about water safety (AwwaRF, 2005).

As described previously (Section 3.2), trust is a central facet of successful communications. Organizations that are visibly present in communities, proactively initiate dialogues, and provide people with information they need and want in a timely fashion are more likely to have successful communication outcomes (AwwaRF, 2004). Trust can also be built by communicating with the public through information sources that the community knows and trusts (e.g., media outlets, community leaders) (AwwaRF, 2004). The style of customer service and communications must be stressed to everyone in the organization (WateReuse Foundation, 2006).

Unique message delivery techniques are still being developed. For example, cognitive scientists at the University of California at Berkeley have recently designed an interactive approach using statistics to assess and restructure people's knowledge about controversial subjects. The approach has been used in a classroom-style setting and online using a software tool known as *Convince Me* (www.convinceme.com). In essence, their method involves first asking participants about their beliefs or impressions surrounding the topic and finding out the range of values or statistics that they expect based on their current understanding. For example, people may be asked about their opinions on incarceration rates, their opinion of an "optimal" rate, and their estimate of the current incarceration rate. People are then provided with statistics about the actual rate.

This method has been shown to result in more profound shifts of opinion, compared with simply providing people with statistics. This may be because people identify with an issue, in a sense choosing a side for their opinion (e.g., pro- or anti-) based on their understanding of a need to increase or decrease that activity. If people shift to thinking about the issue in absolute terms (actual numbers), they may find they are on the opposite side of an issue. Or they may realize that people who take a pro- stance may actually prefer a lower level of that activity than someone with an anti- stance, due to their different perceptions of the current level of that activity (Joseph, 2008).

3.5.3.5 Effective Spokespersons

Spokespersons appear more trustworthy when they have certain qualities, including knowledge and expertise, empathy and caring, honesty and openness, and dedication and commitment (ATSDR, 2004). Knowledge and expertise may be demonstrated by the spokesperson's ease of answering questions and having facts/details on hand, both of which can be improved by experience and prior preparation. Other ways to demonstrate knowledge and expertise include maintaining professional registrations/licenses, publishing in peer-reviewed literature, and collaborating with other reputable organizations or individuals (AwwaRF, 2008a). ATSDR has compiled a number of guidelines (ATSDR, 2004) that support a spokesperson in demonstrating knowledge and expertise. These include maintaining credibility and preparing for public meetings – knowing the audience; anticipating interests, concerns, and questions; practicing responses; preparing the presentation with a strong introduction and three key messages; and other tips (ATSDR, 2004). People are more likely to trust citizen group spokespersons who demonstrate knowledge/ expertise (AwwaRF, 2008a).

Demonstrating care and concern is particularly important for industry spokespersons (AwwaRF, 2008a). Ways to demonstrate these qualities include listening to the public's specific concerns (see ATSDR, 2004 for a list of do's and don'ts of listening based on research by Atwater in 1989); speaking clearly and with compassion; interacting with the community to build trust; and being honest, frank and open (ATSDR, 2004). Many of ATSDR's "ten deadly sins of communication" speak to the importance of seeming knowledgeable and demonstrating care and concern (ATSDR, 2004):

1. appearing unprepared
2. handling questions improperly
3. apologizing for yourself or the organization
4. not knowing knowable information
5. unprofessional use of audiovisual aids
6. seeming to be off schedule
7. not involving participants
8. not establishing rapport
9. appearing disorganized
10. providing the wrong content

ASTDR also provides a list of pitfalls for spokespersons to avoid, including the following: use of jargon; humor (unless used at yourself); referencing negative allegations or negative words and phrases; relying on words only versus visuals; losing your temper; assuming that you have been understood; using abstractions, body language, and other nonverbal messages that are inconsistent with the main message; attacking or blaming other people or organizations; making promises that are not kept; discussing guarantees; speculating; discussing money spent on the issue as a way to indicate concern; making "off the record" remarks; discussing risk-benefit or cost-benefit analyses; using irrelevant risk comparisons and others (ATSDR, 2004).

Another frequently encountered piece of advice for spokespersons is to stay focused on the key messages. These messages (typically a maximum of three) should be the focus of any presentations that the spokesperson gives, followed by information that supports these messages. Use the question and answer period as an opportunity to reiterate key messages. ATSDR provides a list of sample questions from the public or the media and guidelines for answering them in a way that reiterates key messages and avoids common communication pitfalls (ASTDR, 2004). Finally, ASTDR recommends that speakers assess their effectiveness at the end of each meeting or portion of the program (ATSDR, 2004).

3.5.3.6 Communicating with the Media

Utilities and other organizations may first wish to consider when it is appropriate to communicate with the media about risks and conduct risk management activities. In a recent study (AwwaRF, 2004), researchers developed a model and analyzed historical communications to determine whether utility communication was correlated with a number of different factors. They analyzed past communications about several different compounds/issues and considered factors such as the type of contaminant, whether it had been detected, the proximity of the issue (e.g., whether the issue was close to home or perceived as happening far away), the frequency of attention to the issue by the media, and dreaded aspects of the issue – both the level of concern raised by the issue and the involvement of vulnerable populations such as children and pregnant women (AwwaRF, 2004). When scoring an issue against these categories, a list of diagnostic questions was used, as shown in Table 5-1 of the AwwaRF report. A neural network approach was used to develop a decision-making tool for utilities. The tool was tested on 104 different contaminants. The two most important attributes for predicting utility communications with the media were determined to be the frequency of coverage of the issue in the media and the impact that the issue had on vulnerable populations (AwwaRF, 2004).

If utilities decide to communicate with the media or with the general public on a particular issue, it is recommended that they meet early and often with the media to increase reporters' understanding of the issues and develop a network connection with reporters, who are then more likely to call them when they need information (WateReuse Foundation, 2004a). When communicating with the media about water in non-crisis times, utilities must understand that water issues are not that exciting of a topic for the general public and will likely be used as to fill news space when other issues are not more pressing (AwwaRF, 2008a).

It is helpful for utilities to prepare information in a format that meets the media's needs, including press releases that addresses who, what, when, where, why, and how and quotes/succinct summaries that focus on the key messages. ATSDR recommends preparing a 10-12 word sound bite for broadcast media and a one - three line quote that delivers the key message (ATSDR, 2004). Finally, utilities are advised to also develop methods to get their messages out to the public directly, rather than using the media (AwwaRF, 2008a). This might include advertisements, brochures, water bill inserts, website information, youth education programs, annual awards, tabling at community events, treatment plant tours or other actions.

3.5.3.7 Smooth Implementation (Water Reuse Projects)

Smooth implementation of a project is critical to achieving and maintaining customer trust. The WateReuse Foundation completed a report that emphasized the importance of smooth implementation of water reuse projects, paying attention to potential snags (WateReuse Foundation, 2006). Some suggestions include:

- getting customers connected to the recycled water system in a process that is free of red tape and bureaucracy;
- developing a customer service program that requires as little as possible from recycled water customers;
- making sure that customers are informed of any recycled water price increases well in advance so they can budget accordingly;
- communicating on a day-to-day basis during construction with commuters, neighborhoods, schools, day-care facilities, hospitals, and other sensitive areas;
- following up on questions and concerns to build trust with stakeholders;
- assigning staff to recycled water operations and conducting staff training in other departments to ensure a full understanding of the utility's new service, including water and wastewater treatment plant operators; and
- meeting with potential customers and discussing the project benefits, rather than just assuming they will be customers once recycled water is available.

Another report emphasized the importance of constant communication during project implementation and after the project was completed (WateReuse Foundation, 2004a). Finally, the WateReuse Foundation offered the advice of persevering and recognizing that every project has obstacles and setbacks (WateReuse Foundation, 2006). Case studies presented in Section 3.5.4 each encountered difficulties but are now providing recycled water to their communities (WateReuse Foundation, 2006). Similar advice may apply to other types of projects where utilities are interacting with the public.

3.5.3.8 Guidelines for High Concern/Low Trust Situations

High-concern, low-trust situations or hostile situations can arise, particularly in response to issues concerning health and the environment. Discussions can be dominated by a few people, often with extreme competitive and antagonistic views (WERF, 2003). Guidelines for communicating in high-concern, low-trust situations have been published by Vincent Covello (see, e.g., *Core Slides: Communicating in high concern/low trust situations: Applications to water security* (Covello, 2006); *Core slides: Risk communication* (Covello, 2007); *Effective Media Communication during Public Emergencies: A WHO Handbook*, prepared by Covello and Hyer (WHO, 2007)). A summary of several principles for dealing with these situations is provided in this section.

The WHO handbook was prepared specifically for public health officials dealing with communication during emergencies. At these times, effective communication can rally support, assure the public, provide them with much needed information, and help save lives. Poor communication can erode public support, distress the public emotionally, lead to additional social, economic and legal costs, and other negative aspects (WHO, 2007). The book contains a number of communication principles and lists of do's and don'ts (WHO, 2007). Some of the same steps are recommended for communication planning (referred to as message mapping by WHO):

- Identify stakeholders for a specified emergency, crisis, or disaster event.
- Identify stakeholder questions and concerns.
- Identify common sets of concerns.
- Develop key messages.
- Develop supporting information.
- Conduct testing.
- Plan for delivery.

In high-stress/low-trust environments, important communication principles include the following (Covello, 2006):

- ***Compassion, conviction, and optimism**:* Covello recommends approaching all communications with the principle CCO: compassion, conviction, and optimism. When people are stressed and upset, they want to know that others care. This is more important to them than what others know. Listening, caring, empathy, and compassion are typically assessed in the first 30 seconds. Spokespersons must also be aware of negative dominance: In high stress situations, it takes on average three positive statements to balance one negative in peoples' minds.
- ***Simplify***: When people are stressed and upset, they often have difficulty hearing, understanding and remembering more than three messages at a time. Their attention span

also decreases. Covello presents a rule of thumb known as 27/9/3, referring to the need to be able to summarize 3 key messages in 9 seconds using 27 words. In high stress situations, people's brains do not necessarily process information in a linear matter. In addition, the number of messages they can take in is reduced from 7 (+/- 2) to 3 (+/- 2).

- ***Message content***: In his presentation, Covello provided examples of the three messages referred to in the 27/9/3 rule. For example, the messages could focus on what people should know, do, and go for more information. Alternatively, the messages could speak to trust, benefits/fairness, and control of the situation, or could appeal to logic, authority, and emotion.
- ***Preparation***: Covello recommends that a spokesperson a*nticipate, prepare, and practice.* A spokesperson can be asked factual questions, emotionally-charged questions, or challenge/trap/pitfall questions and must be prepared to answer them in a way that reiterates the key messages. Using the phrases "I don't know but [key message]…," "we are doing [key message]…," and "we do know the following: [key message]…" can help spokespersons redirect the focus of questioning back to the key points.

These tips are consistent with ATSDR's recommendations for managing hostile situations – to be prepared, communicate empathy and caring, and stay on track with key messages. ATSDR also recommended acknowledging the presence of hostility and practicing self-management (ATSDR, 2004). This is echoed by a WateReuse Foundation study which recommends that spokespersons practice conflict resolution in preparation and do not get defensive under high-stress, low-trust, situations (WateReuse Foundation, 2004a).

Nonviolent communication (NVC) techniques may also be helpful to utility spokespersons in effectively dealing with hostile or low-trust situations with other stakeholder(s). NVC techniques were developed by Marshall Rosenberg to support compassionate giving and receiving to meet the needs of all concerned. The technique emphasizes compassion as a motivation for action rather than fear, guilt, shame, blame, coercion, threat or punishment (www.cnvc.org). NVC involves empathic listening and honestly expressing through the four steps of the model: observations, feelings, needs, and requests.

The NVC process involves: 1) observing what is happening in a way that is free of evaluation and specifying behaviors and conditions that are affecting us; 2) being able to identify and express internal feeling states in a way that does not imply judgment, criticism, or blame/punishment; 3) connecting these feelings with underlying universal needs/values (e.g., trust, understanding) that are or are not being met in relation to what is happening and how we are feeling; and 4) requesting what we would like in a way that clearly and specifically states what we do want (rather than what we don't want), and that is truly a request and not a demand (i.e. attempting to motivate, however subtly, out of fear, guilt, shame, obligation, etc. rather than out of willingness and compassionate giving) (www.cnvc.org).

For example, in responding to a hypothetical situation of an irate stakeholder, a spokesperson might: 1) observe that the other person is speaking in a raised tone of voice, has left their seat and is questioning the utility's true motives for a project; and 2) notice that internally, he or she is starting to feel afraid, overwhelmed or angry. These feelings might be connected to an underlying need to feel safe or not be intimidated during the conversation, a need to be heard or perhaps a need to be praised for working long and hard on a difficult project.

A request to have his or her needs met would then follow and could be communicated to the other person. For example, the spokesperson might respond to the irate stakeholder: "You have left your seat and are leaning over me, talking in a loud voice. I am starting to feel unsafe. Can you please return to your seat and give me a moment to think about what you are saying?" Although a spokesperson may never be put in this difficult situation, one of ATSDR's guiding principles for a spokesperson in a crisis situation is to plan now; don't just hope that a crisis never comes (ATSDR, 2004).

3.5.4 Utility Communication Programs

Several case studies of utilities conducting water recycling were presented in a report by the WateReuse Association (2004), including a brief description of some utilities' communication programs. Many utilities, particularly the smaller ones, lack the resources needed to host a continual and effective public communications campaign on a variety of issues, including trace organic compounds (AwwaRF, 2008a). However, some utilities, including the ones highlighted here that are practicing water reuse, have existing programs in place and conduct outreach to the public about water recycling (WateReuse Association, 2004).

St. Petersburg, Florida conducted educational outreach, including TV broadcasts, online information, annual awards, community events, and youth education program. At Irvine Ranch Water District, California, there is an extensive public education and involvement program that has been going on for the past 30 years. IRWD has found that a well-informed public is less apprehensive about water reuse. In Orange County, Florida, the Water Conserve II project began as a cooperative water reuse project by the city, county, and agricultural community. The project grew out of a litigation resolution regarding discharge of treated wastewater to a creek. West Basin Municipal Water District, California has a public outreach program, including children's education program, marketing, conservation, and school education. Local cities, community groups, and environmental groups are incorporated into outreach activities.

Water recycling projects in the Salinas Valley (Monterey County, California) were begun to counteract seawater intrusion. Monterey conducted studies of produce grown using recycled water, studies of field worker health, and a marketing study. An independent laboratory was hired to conduct a study on recycled water and food safety. In 1996, a committee was formed to address growers' concerns and get their input. In 1997, a proactive education plan was developed to address perception issues and materials prepared for local growers and sellers in anticipation of questions from the public. Local media and public were reached by offering tours, presentations, billing inserts, and school programs.

This information was published in 2004. More recently, interviews with ten different utilities were conducted as part of this project. Several interview questions focused on the utilities' existing communications program and outreach materials on the issue of trace organic compounds. Results are presented in Section 4.

3.6 Lessons Learned from Other Industries

Under the direction of Task 2 advisor Dr. Paul Slovic, a selection of articles on risk communication about nuclear power, chemical manufacturing, pharmaceuticals and food production, and on the general topic of risk communication were identified and reviewed. The purpose of this review was to identify and illustrate effective communication principles and/or lessons learned from the failure of risk communication attempts. Both the nuclear and chemical industries have faced serious public opposition in the past. The pharmaceuticals industry may be perceived as less of a threat to public health because the benefits of medications are perceived to be strong and the industry takes communications and marketing (product image) seriously. All three industries address trace levels of contaminants that the public is inadvertently exposed to. Although these industries are for-profit organizations, the same communication principles apply to municipalities such as water and wastewater utilities. Six guiding principles or lessons learned from these studies include the following:

1. Pay attention to risk communication.
2. Build trust.
3. Poll perceptions before designing a public relations campaign and during the campaign.
4. Emphasize benefits.
5. Adopt inclusive decision-making processes.
6. Recognize the role and implications of stigma.

The six findings underscore many of the same communication principles identified during the focused literature review of communications in the water industry (Sections 3.1 through 3.5). Each key finding is described in this section.

3.6.1 Lesson #1: Pay Attention to Risk Communication

In his article "Three Mile Island – 25 Years Later," Peter Sandman describes one factor that worsened public perceptions of the nuclear power plant malfunction at Three Mile Island in 1979: risk communication professionals were not substantially involved or included in communications during and after the event (Sandman, 2004). The principal spokesperson, who was the engineering vice president of Metropolitan Edison, viewed risk communication and public relations (PR) as unimportant and had been quoted as saying, "PR isn't a real field. It's not like engineering. Anyone can do it" (Sandman, 2004).

Although there were no deaths and no long-term public health impacts observed or predicted from the doses released during the event, the incident is widely misremembered as a public health catastrophe (Sandman, 2004). Communication during the event was often garbled, with multiple spokespersons presenting conflicting information and using phrases such as "normal aberration" (TIME magazine, 1979). Spokespeople attempted to make their communications reassuring. However, they understated the real potential for a catastrophic release from the facility, which became apparent later, resulting in a loss of trust. The Three Mile Island incident became a potent symbol of nuclear risks with devastating impacts on the nuclear industry (Sandman, 2004).

Sandman concluded that an organization must take risk communication efforts seriously and support them in a concerted and consistent effort, or else the risk communication program would likely not be effective (Sandman, 2004). In the absence of a well-managed risk communication program, people do not have a source of reliable information that they can trust. People will fill this information gap with their worst fears. Social psychologist Stephen Coslett said of the Three Mile Island incident: "What we lack in fact, we usually make up in fiction, and we usually make it up worse than it is" (Cline, 1979).

According to Leiss (2001), the public has faced an information gap about dioxins since the 1970s. Dioxins were introduced to the public as "by far the most toxic compound known to mankind," "the deadliest poison ever made by man," and "more poisonous than cyanide or strychnine" (Leiss, 2001). The presence of dioxin in Agent Orange herbicides linked it with Vietnam War atrocities and to the severe acute health effects reported by veterans. Many misleading statements about dioxins went unchallenged, and no effort was made by government, scientists or industry to counteract these negative associations or offer any alternative perspective of dioxin health effects in terms that were easily understandable to the general public (Leiss, 2001). Thus, the perception of risk from dioxins is still high today despite a decrease in human health effects from dioxin exposure and a decrease in total dioxin burden to the environment (Leiss, 2001).

When an organization does not take risk communication efforts seriously, their reputation can be tarnished. According to Richard Long, a former Dow Chemical Company (Dow) employee for 24 years and director of communications for six years, Dow did not pay attention to public relations or risk communication from the late 1960s until a change in management in 1984 (Long, 2001). During that time, Dow's name became synonymous with environmental and health controversies, due to several stories reported by the popular press. These included napalm manufacturing during the Vietnam war, Dow's response to remarks made at an anti-war speech at Central Michigan University (which was to withdraw funding from the university), other Dow protests on university campuses, a publicized lawsuit against the U.S. EPA for taking aerial photographs of a Dow facility, product litigation over an anti-nausea drug that was linked to birth defects, Agent Orange herbicide manufacturing, Dow questioning after a Greenpeace ship was sunk, sweetheart relationships with the federal government, soil contamination at a Dow manufacturing facility, and more (Long, 2001). Senior leadership at Dow paid little attention to public protests and media coverage of these incidents. The organization received approximately 15,000 calls from the media per year, which were often answered by managers or technical people rather than public relations staff (Long, 2001).

In his article, Long writes that the company did not pay attention to public opinion for several reasons. They believed it did not affect their sales and profit performance, stock price or stock ownership, their ability to recruit or retain employees, or their relationship with the immediate communities surrounding their manufacturing complexes, where Dow was clearly appreciated as a major employer and financial contributor to community parks, libraries, and other public services (Long, 2001). In addition, the company culture was not prepared for the social and political changes in the 1960s and 1970s and may not have known how to respond differently (Long, 2001). Most of the leaders were engineers and chemists who relied on hard data and tended to ignore the role of emotion and politics. Several management decisions were made out of stubbornness and resistance to public opinion, rather than being guided by it, including the decision to continue manufacturing the registered herbicide 2,4,5-T despite its unpopularity on the market (Long, 2001). Dow ignored poll after poll showing that the chemical industry was second only to the tobacco industry in having low favorability and credibility (Long, 2001).

Dow recognized that the company was impacted by the negative public opinion, however. Senior Dow management were constantly "distracted" by public conflicts; the CEO at one point complained that he spend one-quarter of his time dealing with dioxin, Agent Orange, and other conflicts, time that could not be spent on other aspects of the company (Long, 2001). Litigation costs increased with Agent Orange and dioxin lawsuits, including a federal class action lawsuit regarding veteran exposure to Agent Orange. Two jury research polls in the early to mid-1980s indicated a change in public opinion among World War II veterans, the only group profiled as not likely to be sympathetic to Vietnam veterans' claims of delayed stress syndrome or delayed health effects caused by Agent Orange. As this group's opinion changed and Dow was faced with picking a jury for trial without an "ideal" jury profile, Dow agreed to settle for $180 million (Long, 2001). Finally, changes in senior management style in the mid-1980s led to employee opinion surveys and consideration of the results, which stated that Dow management needed to take responsibility for the company's reputation. Dow's employees highlighted several problems with the company's public relations approach. The change in Dow management in the 1980s resulted in a shift in corporate culture, outreach, and branding campaign (Long, 2001).

3.6.2 Lesson #2: Build Trust

The role of trust in risk communication has been recognized by previous studies spanning several different industries (see e.g., Slovic, 1993; Covello, 1991; AwwaRF, 2008a). It is common sense that communication is easier when trust is present. The topic of how to develop trust is the focal point of risk communication studies. Tips include considering the effects of personal control, transparency, public participation, and several communication principles, as described in the following paragraphs.

In a recent survey of the general U.S. public related to pharmaceuticals, Slovic found a positive correlation between peoples' ratings of trust for different groups and their perception of the groups' abilities to reduce the risks of prescription drugs (Slovic, 2007b). In other words, people who assigned more trust to a group (e.g., Congress, hospitals, and patients) perceived that group as having greater control over risks. This suggests that increasing the groups' perceived ability to control risk could promote trust (Slovic, 2007b) and implies that reassuring messages from these groups would be beneficial for the relationship. However, if people perceive a risk as uncontrolled or worse, uncontrollable, and others tell them that they have those risks under control, trust would decrease. Therefore, the public's underlying perceptions of control and controllability must be well understood so that messages can be better attuned to people's perceptions.

The correlation between trust and control can be used to build public trust. In another public survey based on a hypothetical scenario, the most significant way for nuclear power plant officials to increase trust was to establish a community oversight committee with the power to shut down plant operations at any time if deemed unsafe (Slovic, 2007a). Those surveyed trusted community members as an oversight group and believed that, through this arrangement, this group had the power to control risks associated with the plant. Implementing a community oversight group would therefore increase public trust in officials (Slovic, 2007a).

Some companies build trust by emphasizing transparency, "managing trust by bold openness" (van Ryan, 2003). In the early 1980s, the respected medical supply company Johnson & Johnson was scrutinized by the media when several people died from taking Tylenol capsules that had been tampered with cyanide. Johnson & Johnson reacted quickly to media reports and began to investigate the claims and allegations. They welcomed journalists to attend company meetings about the incident, in essence inviting them to uncover the truth. This forthrightness, in combination with the organization's prompt and thorough investigation, and no findings of company misconduct or inappropriate practices, did not add up to a story with "spin" and the media soon moved on to other topics (van Ryan, 2003).

The concept of transparency and community participation/oversight is frequently used at Superfund sites. A community group or members of the community are invited to participate on an advisory board to make decisions about site investigation and remediation. At some sites, funding is provided to hire a technical advisor who acts on behalf of the community (van Ryan, 2003). The involvement of community members and inclusion throughout the process builds trust as well as personal relationships between members of the community and other stakeholders.

Several communication principles can be used to increase trust. In analyzing the Three Mile Island incident, Sandman's advice was as follows: "*Err on the alarming side,*" "*don't lie and don't tell half-truths,*" and "*pay attention to outrage.*" At Three Mile Island, Sandman wrote that company spokespersons were too reassuring, especially at the beginning when no one knew exactly what was happening inside the reactor. By putting the best face on the situation, company representatives found themselves having to go back to the public and say that the situation was worse than they initially thought. This resulted in a loss of credibility. From the perspective of increasing trust, it is much easier to err on the alarming side (e.g., call for an evacuation just to be on the safe side) and later tell people that the situation was not as bad as initially feared. This approach was taken at Three Mile Island by the Pennsylvania Department of Health, who initially issued warnings against drinking milk until the milk supply could be tested for radioactive iodine. Because the state erred on the side of caution for public safety rather than economic impact, its credibility was intact when they later declared that the milk was safe to drink (Sandman, 2004).

Sandman also recommended not saying things that may be technically accurate but misleading, or statements that are easily misinterpreted. People sense that they are not being leveled with or may end up learning the other half of the story and feeling deceived. Regarding Three Mile Island, one spokesperson said that the reactor was "cooling according to design," referring to the reactor design elements that are meant to keep the reactor intact to survive a serious accident. It was the equivalent of saying that, after a major earthquake, a shattered but still standing building was functioning as intended. Although technically accurate, the statement would be contrary to people's emotions, visual impressions, and the obvious contrast between the damage and their image of design and function during normal non-crisis times.

Sandman also recommends paying attention to outrage. This will help a company or industry identify and target the underlying concerns that are feeding peoples' fears, such as mistrust, know-ability/uncertainty, social or racial inequity, lack of control, dread, and memorability of the event. Pinpointing the components of peoples' fears is helpful before designing a communications approach to counteract them. Examples of ways to counteract outrage include changing communication practices, increasing trust, credibility or equity, reducing the sense of uncertainty, involving the public more, or coming up with ways that the public can have more control over the risk.

3.6.3 Lesson #3: Poll Perceptions before Designing a Public Relations Campaign and During the Campaign

Social scientists are familiar with market research tools such as surveys and focus groups to gauge public opinion, design and package products, and evaluate political campaigns. However, these tools are surprisingly under-utilized by engineers and scientists who interact with the public on a variety of technology risks. Findings can be used to understand people's perceptions of risks and to compare the perceptions of subgroups. For example, in a word association survey conducted by Slovic et al. (2007b) about prescription drugs, people who avoided taking medications were found to have statistically significant differences in perception from the other groups (Slovic, 2007b). Conducting a poll or survey before designing or launching a public relations campaign can save money, identify and focus the campaign on key messages, and gauge or predict the campaign's effectiveness.

An example of a failed advertising campaign that did not follow this practice was described by Flynn et al. (1993). The advertising campaign was launched by the American Nuclear Energy Council (ANEC) to generate more public support of a high level nuclear waste repository in Yucca Mountain, Nevada. Narrated by a former popular sportscaster, the television advertisements were intended to demonstrate the safety of transporting nuclear wastes and dispel common misperceptions about nuclear wastes.

A follow-up study after the campaign was launched indicated that most people who opposed the Yucca Mountain repository did not change their opinion after watching the advertisements – only 3.2% thought the ads had increased their support, compared with 29.3% who felt the campaign had made them less supportive. In fact, the campaign appeared to be counterproductive. People's reactions to the advertisements varied, with more than half the viewers (63.6%) reporting negative reactions, including not believing the advertisements, feeling insulted by them or disagreeing with them for a variety of reasons (Flynn et al., 1993).

The campaign failed because public perceptions were not well characterized or understood before the advertisements were designed and launched. As a result, campaign messages did not address or assuage public concerns. Six weeks after the completion of the survey, the campaign itself became a controversial issue. Documents describing the ANEC's campaign strategies were released. These documents stated that the purpose of ANEC's campaign was to sway public opinion towards acceptance of the repository or towards the belief that the repository was inevitable. The documents described specific target groups that could be "informed, reassured, and moved" by the campaign, and training of Department of Energy (DOE) scientists by campaign managers in an attempt to "convince the public that nuclear energy is safe" (Flynn et al., 1993).

Much expense and damage to the ANEC's reputation could have been avoided if ANEC had more carefully informed the design of advertisements and/or pre-tested the messages using focus groups.

3.6.4 Lesson #4: Emphasize Benefits

The benefits to society from various activities and technologies (e.g., nuclear power, electric power lines) tend to be positively associated in the world, that is, higher risk activities tend to have higher benefits to society (Slovic, 2007a). However, people's ratings of perceived risks and benefits of different activities are typically inversely correlated; higher perceived benefit is associated with lower perceived risk and vice versa (Slovic, 2007a). Alhakami and Slovic found that this inverse relationship was a result of people's reliance on general affective evaluations when making risk/ benefit judgments. If people's overall reaction to the technology or activity was favorable, they rated it as high benefit and low risk; if unfavorable, they rated it as low benefit and high risk (Slovic, 1999).

Since risk and benefit perceptions are a function of people's overall affective reaction, anything that increases the positive value of a technology or activity will likely decrease their perception of risk. Peoples' overall affective reactions are of course a function of individual and societal values (NRC, 1996); not everyone will share the same perceptions or values and will not respond the same way to a message about benefits. Benefits may not be distributed evenly over the population – some may benefit from a technology or activity while others do not. Finally, those who benefit may not be the same as those who pay the costs, leading to social inequity (NRC, 1996). These complications aside, people will react more positively to a technology or activity that clearly benefits them.

The perception of medical benefits appears to have had an impact on perceived risk. A study by Slovic et al. found that allergy medications, asthma drugs, and antibiotics were perceived to be lower in risk than sleeping pills and diet drugs, presumably due to higher perceived medicinal benefits (Slovic et al., 2007b). This study found that many pharmaceuticals were perceived to be high benefit and low risk (Slovic et al., 2007b).

3.6.5 Lesson #5: Adopt Inclusive Decision-Making Processes

People value personal choice and individual autonomy; they are more comfortable accepting risks that are voluntary (Covello, 1991). Therefore, people prefer having some decision-making power regarding a technology or activity involving risk. Under the National Contingency Plan (NCP) and other regulatory frameworks, decision makers are required to solicit public opinion via a comment period and response to comments or a public meeting. However, when the public is merely asked to comment on a project that is almost complete, they are not included in the problem formulation stage (Drew et al., 2003). Focus groups and surveys conducted by Drew et al. (2003) on the topic of nuclear waste transportation and repositories highlighted the need to involve people in a transparent decision-making process, so that people who are interested in the decision can understand what is being done and why, and trust that certain processes will consistently lead to specific decisions (Drew et al., 2003).

Decision makers may hesitate to include citizens in decisions about nuclear waste transportation or other areas involving specialized knowledge or uncertainty. However, there are documented examples of nontechnical people grappling with highly complex issues to affect meaningful change (Drew et al., 2003). In our democratic society, policymakers and other leaders must provide the public with the information they need to participate in the decision-making process (Drew et al., 2003).

As described by van Ryan (2003), public involvement is more extensive for some environmental issues. Several remediation sites have involved public representatives throughout the investigation and remediation process. Full participation in site decisions emphasizes the importance and value of both the decision-making process and outcome, increases transparency, and builds trust. Such an approach is more time-consuming and may be more expensive in the short-term, but results in public support of the final remedial decision and litigation avoidance.

Organizations will also benefit from including their employees in decision-making processes. This may be as simple as keeping staff informed about key issues and soliciting staff input. It may also be accomplished by improving processes for communication between different departments or divisions. Sandman reiterates the importance of this communication between key decision makers and company employees, particularly during a crisis or another event: "Keeping people informed and letting them talk is a wiser strategy than keeping them ignorant and hoping that they won't" (Sandman, 2004). Journalists may contact both official and unofficial sources within the company to prepare a news story. Some journalists prefer to contact employees who are not in the public relations department in order to hear alternative points of view and additional facts that are not part of the "company line." Intra-organization communication benefits technical and public relations departments. Journalists may ask for details (e.g., sampling frequency, process) that public relations spokespersons would not necessarily know unless they had frequent communication with other department personnel. From communicating with public relations staff, technical staff will better understand the format of information that journalists need, common questions to expect, communication principles, and other tips.

From an industry perspective, agreement/endorsement of experts from several different agencies will strengthen key messages. Sandman notes that expert disagreement generates fear and outrage by exacerbating the sense of uncertainty normally associated with risk (Sandman, 2004). For example, in the coverage of the Three Mile Island event, medical experts disagreed about the effects of low levels of radiation exposure (Sandman, 2004). If there is some scientific disagreement, industry-wide consensus statements on the topic can help the public and media identify what issues and facts experts do agree upon and place minority viewpoints in perspective.

3.6.6 Lesson #6: Recognize the Role and Implications of Stigma

Several articles have been published regarding the concept of stigma and implications for risk assessment (see "Risk, Media and Stigma," edited by Flynn et al., 2001). The concept of stigma may be useful in understanding social opposition to technologies, places, and products that are perceived to be dread or dangerous. Stigma is a word derived from ancient Greek to denote a mark on an individual signifying infamy or disgrace (Gregory et al., 1995). The word also has connotations of overturning or destroying a positive condition; tainting something that was or should be good, "invoking deep-seated fears associated with the benign or beneficial suddenly becoming menacing" (Gregory et al., 1995). Stigma is similar to the principle of contagion or contamination: physical contact with the stigmatized object is thought to result in permanent, dose-insensitive transfer of fear or disgust (Rozin, 2001). Stigma is often experienced as an invasion of one's body or living space (Satterfield et al., 2001). Many environmental contaminants evoke this sense of contagion or bodily invasion, resulting in unseen or latent damage and interfering with the pleasure of day-to-day experiences interacting with the world (e.g., sitting outdoors, letting children play in the yard, eating food, and drinking water) (Satterfield et al., 2001).

Other researchers have described stigma as an unwarranted level of avoidance behavior based on some marker (associated with a person, place, product or industry) that triggers a negative affective response. This negative response has already occurred elsewhere and has been transferred to this new context via the marker (Walker, 2001). Hence, the concept of stigma utilizes emotional reactions that were already there. The degree of the response is not fully explained by the "real" risk of the situation (Walker, 2001).

Walker's definition raises several difficulties in using the concept of stigma to better understand and respond to people's reactions. Scientists may think they understand the "real" risks better than the public but instead may focus their attention solely on risks that can be quantified through the risk assessment process, a paradigm that does not necessarily capture all of the "real" risks. "Real" risk, in fact, may not be measureable. Thus, stigma is defined by this judgment of a person's affective response as not fully explained by the scientists' perceived risk. In addition, there is no scale or baseline to gauge appropriate reaction since risk tolerance can vary by individual. For example, gender and racial differences in risk tolerance are well documented, with white males typically less concerned about risks than white females and minorities (Slovic, 1999). The word "stigma" has negative connotations to scientists and is often placed in the same mental category as "superstition," implying irrationality and unwarranted behavior. As long as scientists are sensitive to these potential pitfalls in thinking, the concept of stigma could aid communication between scientists and the public (Walker, 2001).

Peter Sandman, Vincent Covello, and other social scientists have listed a number of factors that appear to be linked to public outrage (Covello, 1991), including whether the risk is natural or anthropogenic, whether children are at risk, whether or not victims can be predicted before harm occurs or identified after harm occurs (i.e., is the harm invisible?), extent of media coverage, trustworthiness of related organizations, and other "aggravating circumstances" (Covello, 1991). These factors were researched in the 1980s to identify how hazards interact with psychological, social, institutional, and cultural processes to amplify or attenuate public response. Amplification or attenuation of risk can occur during the communication process (the transfer of information) or in the societal response. A description of the factors influencing the social amplification of risk was prepared by Kasperson et al. (2003) who coined the term "social amplification of risk framework" (SARF). Slovic compiled a similar list of factors: whether the risk was uncontrollable, dread, abnormal, and whether or not there were available substitutes (Slovic, 2007a).

Kunreuther and Slovic describe strategies for dealing with stigma, to be applied to situations where the social and economic response appears to be unwarranted by the risk and leads to "impacts that are far more serious than the initial threat" (Kunreuther and Slovic, 2001). Their suggestions for reducing people's vulnerability to stigma include the factors listed in Table 3-2.

Table 3-2. Strategies for Coping with Stigma.*

Strategy	Component	Description
Prevent the occurrence of stigmatizing events.	Model stigma impacts as explicit costs in decision analysis.	The authors suggest that preventative measures are not taken due to economic factors. Performing an economic analysis that takes into account the costs of each stigmatizing event is recommended to justify the additional costs of preventative measures.
Reduce perceived risk.	Create and maintain trust.	See Finding #2 above.
	Inform, educate, and desensitize the public.	The authors note that this is peoples' typical initial reaction, but is not always effective. For example, risk communication has been shown to be ineffective in reducing phobias, whereas systematic desensitization or counter-conditioning is effective. However, sometimes education can reduce the public's negative reaction, for example, increasing understanding the body's natural defense and repair mechanisms against carcinogenic substances.
	Educate scientists about how risk studies and quantitative risk assessments breed fear.	If possible, present scientific data from the perspective of a No Observed Adverse Effects Level (NOAEL) model for cancer risk rather than a non-threshold linear model (Purchase and Slovic, 1999).
Reduce the number of stigmatizing messages and their social amplification.	Educate the media about stigma.	Educate the media about the potency of stigma and the impact of media language and framing.
	Educate the government about stigma.	Government methods of risk assessment may contribute to stigma – for example, the terminology for probability to cause cancer assigned to different compounds.
Reduce the impacts of stigma.	Provide insurance.	Environmental insurance is already available for Brownfields sites.
	Guarantee property values.	In some areas, developers have guaranteed home values in select neighborhoods for the first ten years.
	Provide compensation.	Compensation can be beneficial for mildly stigmatizing actions or events (e.g., prison or landfill siting), but not in cases where actions or events become more stigmatizing (e.g., nuclear waste storage).

*Content adapted with permission from Kunreuther and Slovic, 2001.

These findings are useful to keep in mind and may explain people's reactions to trace organic compounds, as described in the report recommendations (Section 6.0).

CHAPTER 4.0

REVIEW AND ANALYSIS OF MEDIA COVERAGE AND UTILITY RESPONSE

4.1 Introduction

Clearly, the subject of media reports on trace organic compounds covering pharmaceuticals and personal care products in water is a very broad topic. This chapter focuses specifically on "health, safety, and environmental risk" communications reported in print and broadcast media. The impact of these media articles were discussed with 10 utilities of differing sizes in various locations. No attempt has been made to evaluate the "risks" of pharmaceuticals and personal care products reported in the media against other everyday risks such as using illegal drugs, automobile accidents, arsenic or mercury levels in water, and even more conventional pollutants such as bacteria and total suspended solids. Media reports in the last few months and especially those reports following the Associated Press article ("AP probe finding drugs in drinking water") published on March 10, 2008, have directed the public's attention toward this issue.

This study has applied a uniform logic to each of the media articles and "utility teaming partner" interviews to standardize, to the extent practical, the impressions the articles gave the reader. The research team then tested whether the communications were creating concerns that the utilities were responding to and addressing in responses, messages, outreach materials, and programs.

The researchers used Google website alerts for the word "water" as a key source of its media-article research domain. Other articles that were provided to them for review during the course of their investigation were also included as described in Section 2. The literature research sources were articles in the common press, not peer-reviewed scientific literature. Therefore, this chapter reports only how the information was conveyed to the public. (The research team is not confirming or denying the accuracy or significance of the findings.) In this chapter, the psychological concept is evaluated qualitatively, extending the empirical work into this important topic. That is, this chapter summarizes the effects of the media reports on how the public might perceive the issue. The research team then spoke to its utility teaming partners to determine their points of view related to media articles in their geographic areas and the public's responses to them. The researchers evaluated whether new programs and approaches resulted from the public's heightened interest in the issue of trace organic compounds in the water environment.

4.1.1 Systematic Approach to Media Article Selection and Review

The detailed approach to the media review was to:

- impose a discipline to the process of selecting and summarizing the media articles;

- represent the findings in an undifferentiated manner free from personal opinion; and
- evaluate relationships between articles that might be obscured by other approaches.

Wording of the articles, terminology, and images were evaluated as a means by which the public might make sense of what they read, including it in their mental model of whether the issue is one of mere interest or one that creates alarm, fear, or distress. The research team looked at the issue of "affect" or whether the language may create a feeling – either positive or negative. Alhakami and Slovic (1994) found that people base their judgments of an activity or a technology not only by what they *think* about it but how they *feel* about it. This information was used to critically analyze the media reports to determine whether words, images, and metaphors that were used in the articles created a sense of risk or concern. That is, were opinions or fears expressed by others presented in the article? This analysis is addressed by coding categories 1, 11, and 15 (see Table 4-2).

Media reports were analyzed with respect to the "availability heuristic" or whether they provided examples of harm that could be readily brought to mind. Did the articles depict actual images or emotional ones that created a greater impact than reading about the science of the issues? (See coding category 11 in Table 4-2).

The researchers looked for information, if any, on the concept of "detection" to determine if it conveyed the notion of something new or ominous or whether it merely reported on the advances in detection methods. Was there an assignment of probabilities or harm noted? Were speculative harms defined or discussed? For example, if the media reports neglected probability considerations, it is possible that the reader might have been led to fixate on the negative outcome – creating anxiety and distress without enhancing understanding. Such reporting could cause the reader to focus on a potential risk rather than on the probability the risk would actually occur. Failure to disclose information was also coded.

For the issue of detection, the research team also looked at how detection was expressed and whether it provided example comparisons to other risks in the environment. They tried to determine if any comparisons were seen as familiar or unfamiliar and if there was any attempt to put the discussions into context. For example, was the issue of scientific controversy or uncertainty discussed? Was an explanation of where things stand (certainty vs. uncertainty) provided and were different views explained? Were impacts on human health vs. the environment differentiated? Was a clearly articulated message or conclusion provided?

4.1.2 Selection of Media Articles for Review and Coding

Twenty-five articles and one cartoon image were selected for review and coding. These articles came from a larger list that CH2M Hill and Malcolm Pirnie collected. In addition, a member of the project advisory team emailed one article to the researchers. Articles, which were in both the national press and the local press, came from the East, West, Midwest, and Texas.

The researchers chose articles that they thought would stimulate interest and motivate individuals to want to understand the meaning of issues related to water quality and water recycling. The selection process involved picking some articles with headlines that involved the talking points related to the issue of pharmaceuticals in water. None of the articles contained headlines or references to trace organic materials. No articles that prominently referenced the term "trace organic materials" were found. The researchers looked for articles that would create some degree of *awareness knowledge* – information that an issue exists. They also looked for headlines that tended to indicate content of *how-to knowledge* or information about how to put the issue in perspective. To do so, the researchers looked for headlines or tag lines such as "what are they doing to you," "does it matter," "are you concerned," and "don't flush, drop them off." The research team also selected articles that they thought might contain some *principles-knowledge* – articles that might contain information explaining the issue in context. (Detailed and lengthy investigative reports and some scientific information are often referenced as principles-knowledge in the public press.)

Finally, samples were selected from a number of media channels: from broadcast media (radio and television) as well as print media (newspapers and magazines). The articles were published between October 2000 and March 2008, with the majority published in 2007 and 2008.

Table 4-1 presents the 26 media sources that were coded and analyzed. Note that sometimes several articles reported information from the same source. A letter superscript in column 4 (source) of Table 4-1 indicates references to the same source. For example, all sources marked with "AP" were keying off the Associated Press national investigative team's five-month inquiry released on March 10, 2008.

Table 4-1. News Sources Coded.

Article	Title	Author	Source	Date	Age
1	U.S. Water Not as Pristine as Thought – Part 1/3 in an AP Special Investigation	Jeff Donn, Martha Mendoza, Justin Pritchard. Associated Press Writers	WCBS Newsradio 880[AP]	3-10-08	Web
2	Animals Show Effects of Drugs in Waterways – Part 2/3 in an AP Special Investigation	Jeff Donn, Martha Mendoza, Justin Pritchard. Associated Press Writers	WCBS Newsradio 880[AP]	3-11-08	Web
3	No Regulations to Keep Pharmaceuticals Out of Drinking Water – Part 3/3 in an AP Special Investigation	Jeff Donn, Martha Mendoza, Justin Pritchard. Associated Press Writers	WCBS Newsradio 880[AP]	3-12-08	Web
4	Are You Concerned about Drinking Water?	"Gwinnett Opinions: What Do You Think?" column, author unknown	The Atlanta Journal-Constitution[AP]	3-26-08	Web
5	Big Pharma Is in the Water	Sarah Rubenstein	The Wall Street Journal[AP]	3-10-08	Web
6	Don't Flush the Pills–Drop Them off	Andra Atteberry	The Benton County Daily Record	4-10-08	Web
7	Drugs Are in the Water. Does It Matter?	Cornelia Dean	The New York Times	4-3-07	Web
8	Facing the Yuck Factor	Peter Friederici	High Country News	9-17-07	Web
9	Rivers of Doubt	Anne Underwood	Newsweek	8-21-07	Web
10	Scientists Probe Gender-bending Chemicals in Potomac Fish	Simon Jennings	Environmental Data Interactive Exchange	1-19-07	Web

Article	Title	Author	Source	Date	Age
11	Testicles Shrinking in Las Vegas Bay	Unknown	Southwest Hydrology [LV]	March/April 2007	31
12	Contaminants Change Fish	Launce Rake	Las Vegas Sun[LV]	11-4-06	Web
13	Flush with Chemicals	Michelle Wallar	The Daily Camera	4-17-05	Web
14	Pharmaceuticals in Waterways Raise Concern	Juliet Eilperin	Washington Post	6-23-05	Web
15	Unused Medicine a "Painful" Waste, Possible Hazard	Kim McGuire	The Denver Post	4-17-05	Web
16	Male Fish Growing Eggs found in Potomac	The Associated Press	MSNBC	12-21-04	Web
17	Pharmaceuticals & Personal Care Products: An Rx for Water Quality Problems?	Gary Pitzer	Western Water	July/August 2004	4-13
18	Endocrine Disruptors-What Are They Doing to You?	Kathy Jesperson	On Tap	Winter 2003	Web
19	Traces of Drugs Found in LV Wash	Unknown	Unknown [LV]	10-16-2000	Web
20	Arlington Won't Identify Pharmaceutical Drug Found in City's Water	Jeff Mosier	The Dallas Morning News[AR]	3-10-08	Web
21	Study Finds Pharmaceutical Drugs in Tap Water	Associated Press	Unknown[AR]	3-10-08	Web
22	Arlington Looks to Eliminate Drugs from Tap Water	Jeff Mosier	The Dallas Morning News[AR]	3-26-08	Web
23	Sewer to Spigot: Recycled Water	Anjali Athavaley	The Wall Street Journal	5-15-08	Web
24	Texas Town Releases Name of Drug Found in Water	Martha Mendoza	Associated Press[AR]	6-9-08	Web
25	Flushing Away Fears	Tom Arrandale	Governing.com	May 2008	Web
26	Editorial Cartoon: Jackpot!	Nate Beeler	The Examiner: San Francisco	3-11-08	12

Superscript Legend:
AP = Associated Press National Investigative Team's five-month inquiry released on March 10, 2008
LV = Location is Las Vegas Bay, Lake Mead
AR = Location is Arlington, Texas

4.1.3 Coding Methodology

A set of coding categories was created based on the literature in risk perception and risk communication. For example, the theoretical model of "social amplification of risk" (Kasperson et al., 2003) centers around the key messages or "signals" emanating from a risk incident or news report of an incident. Therefore, the researchers looked for the signals or themes of each article. For example, is it an emerging problem? Is risk new? Is risk increasing? Is there someone to blame? Are authorities competent and trustworthy? In addition to coding for key messages or signals, they looked for other factors known to contribute to social amplification of risk. These factors include:

- Do the messages or signals lack transparency? For example, is information about the risk being suppressed? Is there a lack of testing?
- Are there limits to the technology being used to identify and control the risk?
- How well do scientists believe they understand the risk?
- Are sensitive populations (pregnant women, children, the elderly, sensitive aquatic species) at risk?

Recent research has identified the importance of imagery and feeling (affect) in guiding risk perception and acceptance (Slovic et al., 2004). Accordingly, articles were coded to determine what images and terms were conveyed. The researchers then determined whether the terms were inflammatory or whether they specified a feeling that the reader was to have about the issue (that is, a concern). They also judged the overall affective tone of the article (ranging from very negative to very positive). In addition, the researchers coded basic elements of risk (including the chemicals named), whether the name was common (e.g., painkillers) or scientific, the scope of contamination, the concentration in water and the likely amount of human exposure to the named chemicals, the conclusions of any risk assessments that were performed, any mention of observable harm, interviews with persons harmed, regulatory actions, and proposed solutions to problems raised in the article. Table 4-2 lists the coding categories.

Table 4-2. Coding Categories.

Coding Category	Description
1	Overall Affective Tone
2	Failure to Disclose Information
3a	Pharmaceuticals Listed
3b	Other Substances Listed
4	Scope of Contamination
5	Concentration and Amount of Exposure
6	Risk Assessments
7	Limits of Control Technology
8	Observable Harm
9	Culpability
10	Regulation
11	Emotional and Inflammatory Images and Terms
12	Sensitive Populations
13	Public Response
14	Solutions
15	Key Messages

After a sample article was coded, project team members Linda Macpherson, Paul Slovic, and Jozee Adamson critiqued the coding. Jozee Adamson then coded the articles under the supervision of Paul Slovic. Whenever possible, the actual text was provided in support of the coded content. For example, in one news report, under the code "scope of contamination" the selected text read: "A vast array of pharmaceuticals have been found in the drinking water supplies of at least 41 million Americans."

The coding proceeded smoothly. In presenting the results, it will be apparent how the coded comments fit each of the coding categories. By presenting the coded text in tables within Appendix A, readers have the opportunity, if they wish, to consider alternative codes or alternative conclusions to those presented. Article numbers in the text refer to the numbered source articles in Table 4-1.

4.2 Media Review

4.2.1 Overall Affective Tone

The overall affective tone of the article, as judged by the coders, is presented in Table A-1. It is clear that the majority of articles (17 articles) were negative in tone, four being very negative. Only three articles were positive. Three articles were mixed or balanced and two articles were factual.

4.2.2 Failure to Disclose Information

About a third of the articles reported a failure to disclose information about chemicals in water (Table A-2). In some cases (articles 1, 3, 5, 20, 21, 22, and 24), officials decided not to disclose findings on the grounds that they might unnecessarily frighten the public (e.g., "…the public doesn't know how to interpret the information"). In one case (Article 20), a researcher asserted that his findings were suppressed, a claim that was denied by an official. Some information was withheld for security reasons (articles 1, 20, 21, and 22), though one report (Article 24) argued that there was no proof that such disclosure would help terrorists contaminate water supplies.

4.2.3 Pharmaceuticals and Other Substances Listed

Fifteen of the 26 media pieces listed specific types of pharmaceuticals found in water supplies (Tables C-3 and C-4). Among the more frequent listings were painkillers, tranquilizers, and antibiotics. Hormones were also listed frequently under various names (e.g., sex hormones, birth control pills, natural and synthetic hormones). Typically, common names were used, rather than scientific names.

In addition to pharmaceuticals, a number of other chemicals and substances were listed in about a quarter of the articles (Table A-5). These chemicals and substances included pesticides, industrial chemicals, caffeine, and chicken waste from poultry farms.

4.2.4 Scope of Contamination

Almost every media article gave some indication of the scope of "contamination" in terms of the geographic location of affected water supplies and the number of people exposed (Table A-6). Sometimes specific cities were named (e.g., Las Vegas, San Francisco, Philadelphia, and Washington, D.C.). Sometimes regions were specified (e.g., Northwest Arkansas, Southern Nevada), or even the entire United Sates. Occasionally, specific rivers or lakes were named. Several articles (articles 1, 2, and 17) noted problems in other countries (e.g., Australia, China) and worldwide. Numerous reports (articles 1, 12, 18, 20, and 24) stated that millions of people are exposed.

4.2.5 Concentration and Amount of Exposure

About a third of the source articles mentioned the concentration or amount of the exposure to a pharmaceutical in water (Table A-7). Typically, the concentrations were said to be small – "tiny" (Article 21), "low" (Article 16), "miniscule" (Article 12), or "exceedingly small" (Article 17). And, when concentrations were given, they were listed as parts per billion or smaller (articles 1, 17, 21, and 24) and put in perspective (e.g., "far below the levels of a medical dose," "consume 17,000 gallons to get the amount in one pill," "a grain of salt in a swimming pool," "one drop in 13 million gallons," "a drop of ink in a 12-million-gallon reservoir"). Note that, in some cases, these types of perspective-giving comparisons may be ineffective (Rozin et al., 2000). These comparisons suggest a known amount of contaminant that people want removed (Eriksen, 1990).

4.2.6 Risk Assessments

About 80 percent of the articles assessed risk in some way (Table A-8). None of these assessments was quantitative. One source (Article 7) said assessing the risk of combinations of chemicals was a problem of impossible complexity. Risk evaluations took the form of assertions or implications by officials that the water was safe or low in risk (articles 1, 3, 5, 7, 8, 14, 18, 20, 21, 22, 23, 24, and 25). However, there were also statements of concern about risk to people (articles 1, 8, and 12) or to the environment (articles 5, 6, 14, 17, and 18). Several articles presented a mix of assertions about safety (articles 1, 5, 8, 17, 18, and 21). Other articles focused on the uncertainty regarding possible risks to humans or the environment (articles 5 and 13: "We don't really know what the risks are").

4.2.7 Limits of Control Technology

One of the main factors influencing perception and acceptance of risk is the perceived extent to which exposure to harmful substances is being properly managed or controlled. Table A-9 shows that about a third of the articles mentioned some form of challenge to the ability of treatment plants to remove chemicals from water. Most of these articles commented on the fact that not all chemicals can be removed by treatment (articles 5, 8, 13, 17, and 21). Some articles noted that only trace amounts of drugs remain after treatment (articles 2 and 17). Some articles (e.g., Article 17) noted that certain medications (e.g., anti-seizure drugs) "pass through treatment virtually untouched." Other articles said that the degree of removal varies depending on the type of chemical and the specific water treatment applied.

4.2.8 Observable Harm

Almost every article attributed some specific observed harm to humans or aquatic organisms (Table A-10). Effects on human embryonic kidney cells, blood cells, and breast cancer cells were associated with small amounts of medication (Article 1). Links between home cleaning and industrial chemicals and developmental disabilities and hypothyroidism were noted (Article 18), as well as some increases in cancer and early puberty that might be associated with "hormone-disrupting chemicals" (Article 18).

The majority of harms described in these articles were to a wide range of aquatic organisms (fish, mussels, hydras). Typically, these harms were sex-abnormalities, which were often attributed to exposure to hormones (articles 1, 2, 3, 5, 6, 7, 8, 9, 10, 11, 12, 13, 14, 18, and 19). Feminization of male fish was often noted in these articles. In several articles, effects on lower organisms or on non-aquatic wildlife were also noted (articles 1, 6, 8, 21, and 23) and highly suspected in others (articles 14 and 18). The term "endocrine disruption" appears in two articles to characterize the riskiness of chemicals in water (articles 8 and 23).

4.2.9 Culpability

Another strong signal in media reports that triggers social amplification (great concern) about risk is the assertion that someone or some company or agency is to blame for unwanted contamination or exposure. Table A-11 shows that about half of the articles indicated culpability. However, in contrast to the usual story that blames industry for contamination, most of these stories put the blame on the public (e.g., consumers of pharmaceuticals whose excretions find their way into waterways). In some cases, the articles single out direct flushing of unused medications as the culprit (articles 1, 7, and 17). Animal steroids and antibiotics are another source mentioned (articles 1 and 16), along with industrial discharges and pesticides (Article 18). However, human discharges are, by far, the most discussed cause of unwanted contamination or exposure.

4.2.10 Regulation

About half of the articles discussed issues having to do with U.S. EPA regulation of chemicals, treatment plants, discharges, and so forth. This material, presented in Table A-12, is predominately negative. It points primarily to a lack of regulation or standards for pharmaceutical products in water (articles 1, 2, 7, 8, 15, 20, 21, and 24). The articles frequently mention the U.S. EPA as an agency that is not adequately regulating chemicals in water. One article (Article 1) mentioned how few providers were testing drinking water.

4.2.11 Emotional and Inflammatory Images and Terms

About a third of the news reports contained images or phrases that would likely trigger a strong negative reaction in people on an emotional level. A selection from Table A-13 includes:

- the situation is undoubtedly worse than suggested (Article 1)
- poison (Article 1)
- hamper reproduction (Article 1)
- damage the brain (Article 1)
- allow human germs to mutate into more dangerous forms (Article 1)
- unique danger (Article 1)
- inadvertent drug users – "a case of hidden mass medication of the unsuspecting public" (Article 7)
- drink out of the toilet (Article 8)
- "toilet to tap" (Article 8)
- something fishy in the nation's water supply (Article 9)
- learning disabilities (Article 18)
- ADHD (attention deficit hyperactivity disorder) (Article 18)
- purifying wastewater so that residents can drink it (Article 23)
- drink out of the toilet (Article 25)

As noted previously, the term "endocrine disruptor" also appears in two articles.

4.2.12 Sensitive Populations

Only two of the articles expressly mentioned greater sensitivity to chemicals among certain individuals. Specifically, Article 1 noted that: "Pregnant women, the elderly, and the very ill might be more sensitive."

Article 18 observed that: "Developing embryos and the very young are thought to be at the greatest risk because hormone activity guides the growth of a baby's nervous and immune systems and programs organs and tissues, such as the liver, blood, kidneys, and muscles, so they will function properly. During early development, this regulation of growth and development is critical for a child."

4.2.13 Public Response

Only four articles contained material on potential or actual public responses to unwanted chemicals in the water supply (Table A-14). Article 7 highlights feelings of dread and aversion to hazards seen as new, invisible, and artificial.

Article 23 adds concerns about cost to the factors influencing public acceptance and links acceptance to early political involvement by the public. According to the survey cited in this article, education about the water cycle also helps.

Article 25 links opposition to wastewater recycling to a universal taboo against consuming excrement. It quotes an official who doubts that the public will accept direct reuse for drinking water "anytime in the near future."

4.2.14 Solutions

About half of the articles (Table A-15) mentioned possible solutions to the problem of unwanted chemicals in water using state-of-the-art technologies such as reverse osmosis to effectively purify water (articles 8, 13, 17, 23, and 25). Other articles presented possible ways that the amount of pharmaceuticals in water can be reduced through bacterial cleansing (Article 2); reclaiming unused drugs (articles 3, 6, 7, and 17); public education about drug disposal and informed selection of over-the-counter substances (Article 9); the design of drugs so they break down into safer compounds (Article 3), and the use of home filters (articles 9 and 20).

4.2.15 Key Messages

Risk perception research indicates that, when reading a story having to do with risk, people look for key messages, known as "signals." These signals influence the individual and societal response to risk. The research team examined the 26 articles for such signals (Table A-16).

The most common signal in these messages is that chemicals in our waterways present a new risk to worry about (articles 1, 2, 9, 10, 11, 12, 13, 14, 16, 17, 18, and 26). What is particularly new is the quantity of pharmaceutical chemicals now being detected in our water. Experts appear worried and uncertain about the degree of risk and indicate a need for more technical knowledge (articles 3, 5, 7, 9, 13, 16, and 17). The possible risks are portrayed as unfair and uncontrollable (articles 1 and 21), not well regulated or managed (articles 3, 12, and 14), and portending long-term consequences to human health (Article 1). New programs and solutions exist to address the problem (articles 6, 8, 22, and 23). However, they face an uphill battle for acceptance (Article 23), in part because of the public's perception of risk (articles 1 and 23), in particular the "yuck factor" (Article 25).

4.2.16 Media Summary and Conclusions

The sample of media articles reviewed is neither a comprehensive nor a representative selection of U.S. coverage of the issue of pharmaceuticals in water. Nevertheless, the research team believes some clear messages in these articles are worth considering with regard to communicating with the public.

The media has always felt a "duty to warn the public" about impending problems. This theme is clearly evident in the articles reviewed. It is assumed that the media cover these topics to supply people with the information they need to make informed decisions about the issue of pharmaceuticals and personal care products in water. One hypothesis is that the way the issue of trace organic compounds/pharmaceuticals and personal care products in water is being communicated in the popular press might greatly alarm people–without giving them useful information for making decisions based on context and priority.

Basically, the articles characterize the detection of trace organic compounds (mostly pharmaceuticals) in the nation's water supplies as an emerging problem that needs to be monitored, evaluated, and managed. They point to findings of sex-related changes in fish that create an unsettling feeling and send a signal of potential serious consequences to humans. These articles point to evidence of harm in fish and emphasize the lack of scientific knowledge regarding the implications of these findings for human health. Other main messages are that regulatory agencies are not doing an adequate job of testing for and controlling this unwanted source of chemical exposure, most of which is caused by the massive amounts of pharmaceuticals used and excreted by people.

The dominant characteristic of a "new risk" are messages that indicate the item is uncontrollable, creates involuntary exposure with dread consequences, and is poorly regulated by the authorities. A new risk message is the recipe for massive public concern, such as was seen in the Alar scare in 1989. At that time, a CBS news program conveyed the notion of a new risk messages about Alar, a growth-regulating chemical (the trade name for daminozide). This new risk message led to the overnight collapse of the apple industry, causing growers to lose more than $100 million loss, despite the lack of firm evidence of harm or risk of harm (Rosen, 1990).

A few articles may attempt to put the data in perspective, pointing out that the amounts or concentrations of pharmaceuticals in water are extremely small. Such articles use analogies such as "one part per billion or trillion measures." These analogies are not likely to have the intended ameliorative effect because of the public's lack of understanding of these measures. Many people lack dose-response sensitivity for exposure to chemicals – if a large dose is bad then small doses are also bad (Kraus et al., 1992).

In addition, few attempts were made to educate the public about the broader issues related to water. These broader issues include dwindling supplies; the nature of the water cycle; the fact that some advanced treatment methods do eliminate pharmaceuticals or other unwanted chemicals; and the global warming/climate change implications of these advanced treatment technologies. Almost no effort has been made to inform the public about modern water treatment techniques that can make water reuse safe and beneficial. A couple of articles provide useful suggestions for keeping pharmaceutical products out of wastewater. However, most of the articles reviewed for this report seem likely to alarm people without clearly pointing to possible solutions to the problems raised.

4.3 Utility Interactions with the Media: Survey of Utility Partners

4.3.1 Utility Partner Questions

Ten utility teaming partners were interviewed, with over 20 individual contributors. (At several utilities, more than one person participated in the interview.) These utilities were located in Arizona, both northern and southern California, Colorado, Florida, and Wisconsin (see Figure 4-1). Some of these states were experiencing water shortages, the majority was considering some sort of reuse program, and three of the utilities had advanced wastewater treatment systems. Each utility teaming partner is referenced anonymously in this summary report. Interviews in their entirety are provided in Appendix B.

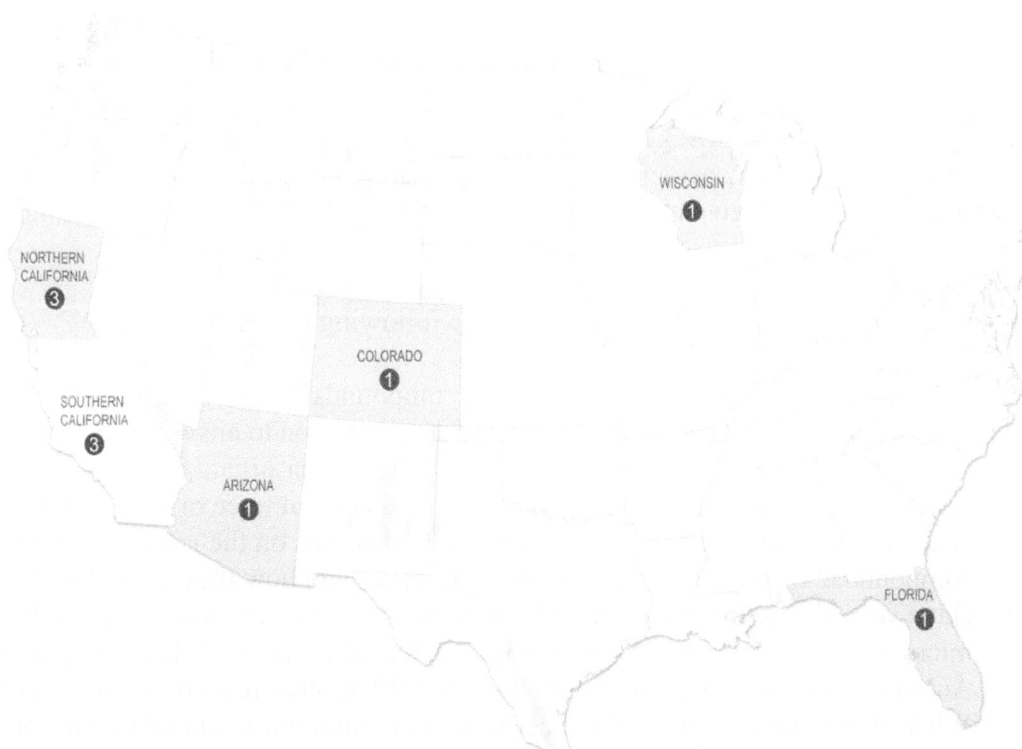

Figure 4-1. Distribution of 10 Utility Teaming Partners Interviewed.
Note: The names of the utilities are anonymous but their geographic distribution is noted generally on the map above.

The interviews were conducted in a disciplined manner. Though the sample size was small and constrained by budget, the interviewers carefully examined the issues. The approach was more analytical than, for instance, the impressionistic analysis of a focus group. Before the interviews were conducted, the project team advisors agreed upon a set of 11 questions to ask each utility. To facilitate discussion, each utility received the questions prior to the interview. These questions guided the topics that were discussed and invited further dialogue.

The interviews were conducted between July 21, 2008, and August 7, 2008. Each interview lasted approximately 1 hour, although a 1.5 hour interval was scheduled for each call to avoid time pressure during the interview. At the beginning of each interview, the interview team verbally updated the utility representatives related to the purpose of the project. They reminded the utility representatives of their critical role, not only in providing input through the utility interview process, but also in reviewing the draft report. They noted that the project scope of work was somewhat narrower than that originally discussed with each of the utility teaming partners. WERF's revised project scope was to conduct: 1) a focused literature review on the knowledge and practice of public perception and risk communication generally; 2) a literature review from risk communication by other industries; 3) an analysis of selected media articles; 4) the input of the utility teaming partners through focused questions; and 5) preparation of a draft and final report. Those interviewed were told that their responses would be reported anonymously in the report and that they would be given a chance to review interview transcript notes and clarify or correct the interpretation of their statements.

Before posing the 11 questions, the interviewers asked each utility what terminology or naming protocols they used. If the utility referred to trace organic compounds as medicines in water, the interviewer replaced "trace organic compounds" with "medicines in water" when asking the questions. Section 4.3.2 summarizes this additional information on the issue of terminology, along with the perceived importance of the topic of trace organic compounds (as it was defined by the utility given the terminology they use). Selected comments from each of the utilities in response to the questions are included in Section 4.3.2.

Each utility teaming partner was asked the following questions:

1. How important of a topic is trace organic compounds within your organization?
2. Does your organization have a designated spokesperson to answer questions from the public about trace organic compounds? Also, does your organization have a designated spokesperson to answer questions from the media about trace organic compounds?
3. Has your organization prepared any outreach materials on the issue of trace organic compounds? If so, please describe these materials and how they were distributed.
4. Have you ever used WEF or WERF's resources on trace organic compounds or microconstituents, such as their website, Technical Practice Updates or Question & Answers? If so, was the information helpful? What other resources have been useful?
5. What, if any, are the principal concerns in your community related to trace organic compounds (e.g., human health concerns, environmental health concerns, other)? Please estimate the percentage of public concern associated with each.
6. Has your organization developed any partnerships to assist in communicating the risks associated with trace organic compounds (e.g., partnerships with local research organizations, medical doctors, consultants)?
7. What tools, processes, and approaches has your organization used to interact with the public? Describe the situations for which each approach has been used. Do you think that these methods attenuate public concern about trace organic compounds?

8. In your geographic area, have there been any changes in practices or programs to address trace organic compounds, such as monitoring programs, treatment studies or pharmaceutical take-back and disposal programs? Were any of these changes a direct result of public concern? Were any initiated by or supported by your organization?
9. Has your organization witnessed any changes in public involvement or perception regarding trace organics over time as a function of setting (e.g., during a drought, in response to an event, in response to the establishment of a program, etc.)? Please describe.
10. Please identify a local media report (article, radio, or television report) as an example of how the media presents information on this issue and how your organization is represented.
11. Has your organization identified any communication deficiencies or needs (e.g., message content, style, trust, choice of media, etc.) that would improve the water industry's communication with the media and with the greater public on trace organic compounds/microconstituents? Similarly, what approaches have been effective? Why?

Verbatim transcripts of the interviews were not made and the interviews were not recorded. Two members of the research team took summary notes, which were compiled at the end of the interview. The utilities interviewed were asked to confirm the summary notes and provide any amplifying information. The utility teaming partners looked hard at the issue. The content that was provided was certainly more than a quick impressionistic analysis.

4.3.2 Terminology and Relative Importance of the Issue

The 10 utilities interviewed did not use the same terminology to describe the issue. Only one utility expressed a strong preference for the term microconstituents, and one liked trace organic compounds. One utility noted that they had not embraced the term microconstituents, while another noted that the public (and the media) do not understand either term, so both should be avoided. That utility used the term pharmaceutical compounds because they thought it was more understandable. Some of the utilities expressed the need for an industry-wide, consistent term, and many felt that it was best simply to use the term that the media uses. Table 4-3 presents a selection of comments about terminology.

Table 4-3. Terminology Used to Convey Messages to the Public and/or Media about the Issue.

Utility	Terminology
1	Endocrine Disrupting Compounds (EDCs) is the term that we have used but sometimes the term emerging contaminants is used. It would be good if the industry developed a consistent term.
2	We use the term "microconstituents" instead of "trace organic compounds."
3	We are <u>not using</u> "trace organic compounds" in communications about this issue. We use various descriptors including "pollutants."
4	It is difficult to pin down one term since there are lots of compounds out there. Sometimes the issue at hand is about pharmaceuticals alone, pharmaceuticals and person care products, or other. The problem with using the term organic compounds or microconstituents is that the public may not understand either term, and the media may not either. You really have to adapt the term you use to what the media is using or feels most comfortable with. There is a limited understanding of the many new media reporters who are just starting to learn about this topic. However, we would be quick to correct a reporter who uses the term "micropollutants" with information that these materials are not in polluting quantities (quantities that would have a negative deleterious effect) and that almost anything could be seen as a pollutant at some concentration.
5	Pharmaceutical compounds is the term that is used because everyone feels they know what that is – microconstituents and trace organics are terms people don't understand. We have also used the term "medicines" because it is more understandable to the broadest range of people and covers both prescription and nonprescription drugs. We are reluctant to use the word "drugs" because of illegal implications.
6	The issue of terminology was not discussed with this utility.
7	The general public is concerned about pharmaceuticals in source waters. The public thinks about pharmaceuticals and does not use the term "trace organics." The phrase "trace organics" relates to volatile organic compounds (VOCs) or disinfection byproducts (DBPs). Neither VOCs nor DBPs has been a public issue.
8	We like the term trace organics but have not embraced the term microconstituents. We feel adaptable to speak in the language of whatever term anyone uses so we have not locked into anything specific. We do not mind using terms that euphemistically evidence "concern." The World Health Organization (WHO) appears to be using the term compounds or contaminates of emerging concern. In a recent meeting of the Singapore Independent Science Panel, Prof Ong Choon Nam and Prof Joan Rose, both international experts, used the term compounds of emerging concern so there may be some growing acceptance though it was not known whether this would carry over to public dialogue where the focus may be different. Within the industry there is more of a tendency to be frank but when we talk to the public there is a need to try to be consistent about language. Almost all of the compounds (except perchlorate) are trace organic compounds. Dr. David Sedlak and Dr. George Tchobanoglous have used the term "pharmaceutically active compounds." The notion of trace organics has been around forever. We have rolled with whatever the current term seems to be and we are comfortable with a lot of terms – haven't tried to push the industry one way or another.
9	We generally use "emerging contaminants" as the preferred term because it is one of the longest used terms. "Compounds of emerging concern" or "compounds of potential concern" are another term that was often used. We suggested the phrase "compounds of potential concern" which was initially embraced by the WEF Community of Practice though the preferred term was changed later. U.S. EPA has used "compounds of emerging concern" for a while. Most people understand these terms more than those currently in use which seem more like semantics or mincing words. It is important that the public has a sense of these items as something of concern but it is true that we don't want to heighten concerns if we don't know the impact of these contaminants.
10	Used the same terminology to describe the compounds as AP (pharmaceuticals, endocrine-disrupting compounds). Did not use one term to define the issue (e.g., trace organic compounds or microconstituents).

Table 4-4. Relative Importance of the Issue as Defined by the Utility.

Relative Importance	Utilities
Very Important	4
Important	5*
Somewhat Important	1
*Will become more important as more detection is noted and detection methods improve.	

4.3.3 Outreach Tools/Approaches/Programs

Table 4-5. Designated Spokesperson for the "Trace Organics" Issue as Defined.

Have a Designated Spokesperson?	Number of Utilities
Yes	4
Handled as a routine communication matter	6
Comments: - Some confusion regarding whether a wastewater spokesperson or a water supply spokesperson should respond. - Varies regarding how technical the question appears to be. - Many are knowledgeable and to some degree every tour guide is a spokesperson on the issue. - Has designated people to respond to the media, but they are not experts on trace organic compounds.	

Table 4-6. Outreach Materials.

Utility	Materials
1	- Has prepared a website and media/public talking points.
2	- No, this utility has not prepared any press releases or other materials on this topic.
3	- Prepared a newsletter which included a general disposal guide though it was not targeted to the issue of "trace organics." - The public may see pharmaceuticals in water the same as fish advisories that are used because of mercury. - Have highlighted triclosan in soaps as a problem.
4	- Has a program in place since mid-2005 for educating people about how to dispose of material including pharmaceuticals. The program won a national award from the National Association of Clean Water Agencies (NACWA). The name of the program is "No Drugs Down the Drain" and it is growing this year to a statewide program. - Since the 2002-2003 timeframe, they have monitored media interest and have been proactive with the pharmaceutical takeback program. It is part of the hazardous household waste program (paints, batteries, etc.) that was adapted to take pharmaceuticals. The program may help people feel that they are part of the solution. - Another example of leadership was the work they did on lindane, an insecticide that has been used in agriculture and also for the treatment of head lice. Lindane is a neurotoxic that is readily skin-absorbed. Led the initiative to get lindane banned.
5	- Yes, the program about medicine take-back was posted on the city's website. They produced several white papers, one on the general subject of trace organics, one on triclosan and one on pharmaceuticals. Has a policy of not using triclosan. They prepared a fact sheet that explains to the public why they decided this and lists soap products that do not contain triclosan. This fact sheet is probably posted on the city's website.
6	- There is an outreach program known as a speaker's bureau program. One topic relates to water quality and would cover this issue.
7	- No specific outreach materials were described but a review of the website noted that there was information about removal provided by the recycling treatment technology.
8	- Have prepared materials that focus on the treatment process and how the multiple barriers remove materials including microbial or emerging contaminants.
9	- A drug take-back campaign as well as highlighted research on triclosan. Proper drug disposal days are noted on calendar listings. Some of the outreach materials have received media coverage.
10	- No specific materials have been put together for the public on trace organic compounds. After the AP article, put together speaking points to be used by people within the city who are receiving calls from the public. Since the interest generated by the AP article, has initiated a drinking water monitoring program and prepared a press release describing the results. No pharmaceuticals were detected, although caffeine was found in one of the treatment plant samples – the one that was not treated by granular activated carbon (GAC).

Table 4-7. Use of WEF and WERF Materials on Trace Organics or Microconstituents.

Use WEF or WERF Materials?	Number of Utilities
Yes	4
No	5
Somewhat	1
Comments: - Info has been helpful. - Would look more to U.S. EPA. - WEF's provision of a link when the AP stories came out was helpful. - Use AWWA and AwwaRF email alerts. - AwwaRF was very active when the AP articles came out. Less attuned to WERF tools. Used other known experts not necessarily those of WEF & WERF. - Some access problems due to membership issues. WEF emails were used. Best sources are individuals and organizations with more subject matter expertise. - The water quality group has not used either the WEF or WERF resources on this topic. The water treatment group may have used these resources.	

Table 4-8. Techniques to Interact with the Public about Trace Organic Compounds.

Utility	Techniques
1	- Returning phone calls - Website - Talking points for city manager - Utility bill insert on drug take-back
2	- Household hazardous waste programs - Responses to phone calls
3	- Not a public issue yet
4	- Has produced brochures, discussed during tours and talks - Has purchased advertisement space in local newspapers - Has a long term drug take-back program
5	- Websites - Brochures - Medicine take-back program - Info about alternative pesticide use
6	- Fact sheet - Website - Speakers bureau with a topic on trace organic compounds (though not by this name)
7	- No formal communication beyond answering calls
8	- Has used multiple techniques from focus groups to surveys - Proactive use of the media to "tell their story first" - Info about how well technology removes pharmaceuticals
9	- Pharmaceutical take-back program - Publicizes information in bill inserts and website - Responds to phone calls - Direct interaction with the public at special events and tours - Provision of outreach materials - Tours which discuss issues such as emerging contaminants (in the context of a plant tour the public more easily gets that it is not easy to change technology) - Collecting unused drugs and preventing pollution
10	- Responds to phone calls - Initiated drinking water monitoring program - Prepared press release with results of water monitoring program

Table 4-9. Partnerships with Others to Assist in Communication About the Issue.

Utility	Partnerships with Others
1	- Have not instituted any partnerships but will partner with a hospital to do pharmaceutical take-back
2	- No partnerships
3	- Partnered with state and county dental societies regarding dental amalgam separators as mercury is a known pollutant
4	- Yes, partnering currently on a statewide drug take-back program - Collaborates with a water authority, U.S. Geological Survey (USGS), several universities - More coordination with scientists than public - Some partnering (review) with media
5	- Coordinated with a local medical association chapter - Worked with nearby water districts and associations - Partner with those focused on pollution prevention
6	- May partner with an adjacent water district or research associations as the opportunity arises - WERF and AwwaRF - Universities
7	- No partnerships
8	- Quite a number of partnership venues including scientific advisory panels - Public is seen as a partner as are health organizations and state regulators along with a watershed authority
9	- Has partnered with two nonprofits especially regarding pharmaceutical take-backs - Partners with a regional monitoring program - Plan to monitor for pharmaceuticals
10	- Participated in a meeting with other cities, the Department of Health Services, and representatives from the county to talk about ways to communicate with the public and discuss key messages - Has worked collaboratively with another city regarding other aspects of trace organic compounds

4.3.4 Perceived Concerns

Table 4-10. Perceived Impact of the Media on Utility Operations.

Utility	Perceived Media Impact
1	March 2008 article in local newspaper created initial concern.
2	No. The issue hasn't come up.
3	One or two articles in local newspapers after AP articles. Sent out an article in response to discuss water treatment.
4	A news article by a local reporter with detailed background helped put the issue in perspective.
5	Hard to remember. Talked with local papers after the AP stories. The local media was pretty good and did not appear motivated to scare.
6	None identified.
7	AP articles didn't make specific mention of this utility. It will take U.S. EPA a long time to promulgate rules after any health effects are known.
8	Media is in constant contact. Tours are helpful with media. Some media representatives like to play on the "yuck factor." This has received print and broadcast coverage. Most reporters have understood media materials provided. Other more sensational stories have pushed this issue off the front page.
9	AP articles were a great example of media coverage but there wasn't a lot of local information incorporated which left agencies scrambling to respond. Some of the AP's facts were wrong. AP reporters should have worked with local utilities first.
10	Released a press statement about the results of drinking water testing in response to the AP article. There has been little public concern since that time. There is no specific information about this issue in the consumer confidence report (CCR) or on their website. Currently thinking about whether to report the drinking water test results in next year's CCR. If they receive more public phone calls on this issue, would consider adding a website reference.

Table 4-11. Principal Concerns – Human Health, Environment (Aquatic Health), or Both.

Utility	Health	Environment	Principal Concerns
1	X		- Drinking water. - Findings on fish could translate into a human health concern.
2		X	- The major concern is environmental. - Deep well injection diminished public concern of all types.
3	X		- Concern focus on whether the plant removes pharmaceuticals and how to dispose of pharmaceuticals. - People do not equate the issue to fish health but worry about eating fish.
4	X	X	- A number of concerns – including environmental concerns. - Those interested in recycled water have focused on human health.
5	X		- Most are worried about drinking water and most don't know where their water comes from.
6	X		- Most of the concern 95% relates to human health issues but environmental groups are concerned about fish.
7	X		- Almost all express human health as a concern. People don't want to be involuntarily exposed to risk. - Environmental concerns have not been raised.
8	X		- More concern about human health. - Pretty sharp awareness of environmental issues.
9	X		- A lot of confusion among residents about the issue. - People want to be told what to buy. - There is confusion about whether the presence of drugs will harm human health. - Since the compounds are known to harm critters, people make the leap that they will harm people.
10	X		- Based on the few calls received, the public was concerned with human health. Some were also concerned about the health of pets.

Table 4-12. Changes in Practices/Programs That Are or Will Be Undertaken Relative to the Concern.

Utility	Changes in Practices/Programs
1	- Will be developing a drug take-back program which will not solve the problem but will help and is something the public can do. - The program above will produce frequently asked questions (FAQs) and media outreach.
2	- No changes in any programs related to this issue.
3	- Thinking about a pharmaceutical take-back program but realizes that the issue is because of urine and fecal matter not primarily because people are flushing unused drugs down the drain.
4	- Already has a pharmaceutical take-back program and continue to participate in studies. - Developing internal capabilities to monitor at the lab. - Permit suggests "special studies" on impact of EDCs. - Observed that there is now a statewide drug take back program.
5	- The state may require some changes and there may be requirements to monitor.
6	- There was reaction to the AP story.
7	- A pharmaceutical take-back program was added to the hazardous waste collection. - Concerned that voluntary testing for pharmaceuticals by an adjacent jurisdiction may create pressure to do sampling which could place people in fear or steer them toward bottled water (which hasn't been tested for these compounds).
8	- Involved in proper solid waste disposal and is involved in the No Drugs Down the Drain Program, but in a support role. - Monitoring is required by permit.
9	- Added a position recently to do more outreach. - Will discuss the issue of nanoparticles.
10	- There have not been any changes in treatment because of trace organics. There has not been any evident increase in public concern about pharmaceuticals other than the response to the AP article. - Most conversations were about testing and testing protocols. - Some adjacent cities have been more proactive and have started pharmaceutical take-back programs. They have not started a take-back program.

4.3.5 Major Deficiencies/Needs

Table 4-13. Identified Communication Deficiencies that Would Improve Media and Public Understanding.

Utility	Identified Communication Deficiencies
1	- Education has been effective and is still needed but there is a need for more public education to help media and public to have enough relevant knowledge to put the issue in context. Ideally this information should come from a national source like WEF and WERF. Some info on the topic is being used by special interests such as religious organizations who want to ban birth control.
2	- Issue has not come up but in two years more communication will be needed.
3	- Deficiencies in accuracy are a part of working with the media in an era of shrinking resources and fewer investigative reporting. Limited scientific coverage.
4	- Need simple message points and a list of tips for how to deal with the media. Need someone to put detection in context. What do trace amounts mean in terms of harm? WEF, AWWA should help deal with the media. WERF has lots of info but the public can't deal with technical jargon. Need an outside organization to help enhance credibility (wastewater utilities and associations may be seen to have a conflict of interest). When wastewater utilities and public health professionals aren't saying the same thing confidence is eroded. The public wants to know why we are learning about this now and may feel something has been hidden. It isn't a new issue. We need more ethical reporting.
5	- The public doesn't understand why there isn't better data. It sounds lame to the public that we haven't tested for various compounds. When you step away from this you quickly see how it sounds silly that we let people take these drugs without having a good testing program about how much gets into water. People can understand that we aren't doing a more holistic job of testing – for example, that we have never tested in the San Francisco Bay for these compounds or worse, have never tested their drinking water.

Utility	Identified Communication Deficiencies
6	- One example is the need to establish safe thresholds for various constituents so that utilities can report to the public that the detected levels are well below or in the same magnitude of the concern. It would be nice to find a way to compare levels of these constituents in water to a safe dose. The medical industry could weigh in if the levels of these compounds are not a health concern. There may need to be more human health research/toxicology studies before we can say that.
7	- The problem we run into is that everything is sensationalized in the media, which creates public fears. People are concerned and start to doubt public agencies and have safety concerns. Getting the good word out about drinking water is difficult. We need to put out an alternative point of view, perhaps through WEF or AWWA as a contact for local news media to restore public confidence, which can be slow. It has taken years to recover public confidence from the cryptosporidium outbreak in Milwaukee. - Another challenge is the lack of scientific knowledge in the general community. It is hard to convince people that water is safe. They may say "why should I trust you? You are not even testing for these constituents. I trust AP more than you as a local politician." They may feel they are just getting a song-and-dance around the issue. Water and wastewater treatment agencies may not be the most credible sources on water quality and we need others' help. WEF and AWWA could help utilities by preparing positive information and messages that could be transmitted to utilities before a media campaign is anticipated to help them with something to say to callers. Topics could include: What are the messages? Who are the experts? - The public needs to know how to put the issue in context. There is more trust in bottled water as a purer source of drinking water than tap water. We have explained to callers that bottled water quality is not regulated and that water is not pure; it is a universal solvent. Some water basics could help the public understand these concepts.
8	- Groups like WERF, AwwaRF, and WateReuse could do a better job of being an objective source of information – we have a long ways to go before we present an integrated water industry perspective. People tend to respond with outrage. The most effective approach has been honest. The first reaction from the public agency and perhaps from the water associations is to cringe. We should be more forthright. - Be prepared – put everyone at ease. Close to having a means to give benchmarks – drinking water equivalent levels (DWELs) if you have a benchmark that says you shouldn't be worried. DWELs went through the toxicological literature – what is a de minimus risk level where we know that the public will find the risk trivial? We need to get to this level so we will know the acceptable level in water supply without causing concern.
9	- It would be helpful if talking points were coordinated between water and wastewater entities. - The industry could also be more proactive instead of only being reactive. For example, a recent article prepared about recycled water for a local newspaper, talked about detections of endocrine disrupting compounds in almost all water supplies, the need to be protective of water, and other key points. The article was very positive because the industry was in the driver's seat for article content. In general the wastewater industry tends to keep their heads low. This does not promote a positive image. If instead, the industry kept telling positive stories, they would balance out the negative press. - In addition, an approach that has really worked is doing regional positive media outreach. A pollution prevention group has members from 43 wastewater agencies and works together on regional projects such as regional thermometer exchanges, regional medicine disposal events, car wash outreach, etc. Again, this is proactive, positive work with the media.
10	- There needs to be more communication with the public to answer their questions related to what it all means, what health studies have been done, what studies are ongoing, and what the risks and dangers are and to articulate what testing results mean. - The meeting with others revealed a lack of experts to communicate with the media on health effects of these compounds, experts who have the necessary background, credibility, and ability to help put the issue in context.

4.3.6 Summary and Conclusions from the Utility Teaming Partner Interviews

The 10 utility teaming partners do not constitute a sample size that is representative of all of the utilities that are grappling with the issue of trace organic compounds or pharmaceuticals and personal care products in the water environment. In addition, those interviewed are not geographically balanced. Nonetheless, those interviewed included representatives of both small and large water and wastewater utilities within both special water utility districts and municipalities. Some of the utility teaming partner respondents were dealing with water supply shortages and may have been more sensitized to the issue of effluent quality because of active non-potable water reuse programs. Some of the utilities were involved in or considering potable reuse programs and changes to technology, while others expressed concern over the costs of monitoring.

From the interview process, the researchers reached the following conclusions:

- An appropriate terminology has not been determined. In general, the media is providing the lead on the topic of how the issue is defined for the utilities and understood by the public.
- Utilities have different beliefs on the issue of trace organic compounds or pharmaceuticals and personal care products in the water environment. On one end, some believe that the issue should be defined as emerging or "a concern." Regardless of the definition, most utilities are treating the issue as a serious problem. However, for at least one utility, the issue was not a concern at all.
- The issue of human health is the driver for the public far more than impacts on aquatic life. Coastal communities have been involved in the issue longer because of the pressure from environmental organizations. If an aquatic health concern exists, a human health concern also exists.
- An entity with credibility (perhaps outside of the water industry) should create communications and public resources to help people understand the issue. It was suggested that the public health professionals should become involved to advance trust because the water and wastewater experts approach this issue differently and they may be perceived as having a self-interest.
- Communication specialists (when they are available) are involved in the issue but technical experts have been responding to the more difficult questions. There has been little follow-up to know whether the public has more, less, or the same concern after speaking with utility representatives.
- The majority of those interviewed are either currently using or planning "drug take back programs" as a response to the issue. Some utilities mentioned that they know this will not change the basic issue of drugs in water. One state is pursuing a statewide drug-take-back strategy.
- It has been determined that even if there is no associated health relationship agreed upon by experts, the issue is one that has created a concern with the public that requires candor and proactive response. There was no determination about what the most appropriate proactive response should be beyond answering the public's questions, providing information on websites, and the selection of a "drug take back program" as a means for community engagement.

- Some felt that the media had sensationalized the issue caused, in some measure, by print media responses to declining revenues. These utility respondents noted both that "sensation sells papers" and that the media has had to let some of the more knowledgeable and experienced reporters go because newspapers are dealing with financial issues. News reporters are not subject-matter experts, and they are encouraged to find the stories that sell the papers. This has a detrimental impact on the utilities because the relationships that the utilities have built have been lost. (One interviewee noted that the issue of drugs in water was front-page news one day, but bumped the next day by revelations in New York about Attorney General Eliot Spitzer, which were of greater interest.) Working with the media proactively was mentioned as one solution to enhancing understanding.

This study was not intended to be representative of all types of utilities but to rather provide examples of utility communication programs and outreach efforts, how utilities are portrayed in the media and how they have addressed the issue of trace organic compounds with the public. To get data that could be analyzed statistically, the questions would need to be repeated with a broader group of approximately 20-30 utilities from various locations. It would also be helpful to ensure that the perspectives of water treatment and wastewater treatment are included evenly in the study and have some perspective on the degree of technology provided. For instance, utilities with advanced treatment, such as reverse osmosis (RO), may be relatively less "concerned" about media attention because their media responses could be different from those plants where trace organic compounds/pharmaceuticals and personal care products were discovered in plant effluent or drinking water supplies and they were unable to assure the public that these materials were removed during treatment.

4.4 Summary and Conclusions

The patterns of media coverage of the issue have helped shape public opinion about the seriousness of the issue. In turn, such coverage may have influenced the response from the water industry. Social research reveals how complicated these issues are. Allan Mazur (1984) examined this issue with four interrelated propositions: 1) a few national news organizations are very influential in selecting which hazards receive most attention each year; 2) public actions to alleviate a hazard or oppose a particular technology rise and fall with the amount of media reporting; 3) public concerns over a risk or technology rise as press and television coverage increases; and 4) it is important to distinguish between the substantive content of a news story and the simple image that the story conveys.

From the responses of our utility teaming partners, the media stories had a dramatic effect in spotlighting the issue. Public interest was generally related to human health impacts. It is likely that the use of images from aquatic studies suggested that risks to people and aquatic organisms are the same. This created heightened concern about reproductive health. This conclusion does not mean that reactions in aquatic species should be dismissed. However, it merely notes that public interest in the issue and the concern that was raised appear to rest, to some degree, on feared but unsubstantiated risks to humans.

The key response to the issue from utilities was to try to help the public by providing information about how they can take responsibility to resolve the concern about pharmaceuticals in water. Product substitutions were one such approach, along with "drug" or "pharmaceutical" take-back programs. Those who have undertaken, or will be undertaking, these programs realize that the issue of detection of pharmaceuticals in water will still exist. However, they believe that some proactive action is needed. Little attempt was made to discuss the issue of detection and the advancement of analytical techniques.

In 1988, researchers (Kasperson, Kasperson, Renn, and Slovic) introduced the social amplification of risk framework (SARF). Two decades later, social amplification of risk perception can be used to examine trace organic compounds/pharmaceuticals or drugs in water as the issue can to the public attention. The images, signs, and signals coded in the media articles, like a stone dropping in water, interacted with a number of psychological and social principles (dread, trust, involuntary exposure, dose response confusion, and inability to distinguish differences between aquatic studies and human health impact). These interactions had a ripple effect that intensified the public's concern. Utilities, in their desire to be proactive on an issue that has created public interest, were concerned about how to respond. They did not appear to want to downplay something that alarmed the public. Only one of the utilities interviewed had had a previous experience of proactively telling the public about an unregulated compound found in their water. That utility now has treatment processes that remove pharmaceuticals from their product water (which is derived from secondary-treated effluent but used advanced treatment to produce water for drinking). All the other utilities either found that the issue did not generate public interest or believed they were providing reactive responses in some way. None of those interviewed felt that they were out ahead of the issue. The desire to be "proactive" was strong. The lack of partnership between the media and the industry was noted. In general, the industry was not seen favorably in the media articles.

The experience of the perceived risk, therefore, is not only the scientific evidence regarding harm but also the result of the media/public perception by which the public has interpreted the risk. The responses that shaped the public interpretation of the risk came principally from media reports. However, as the risk signals cascaded from the media to the public to the utilities, the utilities were left in a responsive or reactive mode. Likewise, politicians were interested in the issue – How did it occur? Why weren't they "told?" Scientists were interested in the issue from an analytical research perspective – "What does it mean?" At each impact, or ripple, the risk could trigger a response to attenuate the risk and put it into perspective against other needed infrastructure investments – including those with known public health impacts (such as mercury in the water). Some utility representatives believed the public began to blur the issues and failed to distinguish between existing programs and this new information. Drug take back programs appeared to be the predominant managerial response to the issue, even though they would yield only a small reduction in concentrations and potential impacts to aquatic organisms. In addition, they would have no demonstrable effect on public health because no risk to public health has been established. However, it was noted by those selecting this program, that it was a good means to get the public involved in resolution of the issue. Additional research will be needed to determine if these programs attenuate or amplify the perception of "risk" associated with trace organic compounds/pharmaceuticals in water. Research will also be needed to determine whether the public perceives the programs will change the detection of these materials.

Did the amount of media coverage generated by the Associated Press articles in March 2008 amplify which water issue would receive the most attention? For instance, during this time the water and wastewater industry was mounting a campaign to focus on the need for more infrastructure investment. Did public interest on this issue push the infrastructure investment need off the agenda? Research will be needed to determine how the public prioritizes water investment issues. In addition, more work is needed to distinguish between the content of the media information and the affective tone that is conveyed. Additional research may be needed to determine if media headlines, cartoons, and artwork created more alarm because they were more memorable. Some of these cartoons had the impact of providing fish with human characteristics.

Other interesting questions relate to how the media are addressing this issue in their investigative reports. Does the media perceive that there is a major health issue that needs to be exposed? The background and the views the media holds might be another fertile ground for research. Did the media consider the water industry as a partner in telling the story or was the industry perceived as an adversary?

Several utilities desired increased effort to create a more balanced framing of the issue through involvement of credible third parties. Better framing of public discourse on the subject could have helped attenuate the risk perception. The interest and energy surrounding drug take back as a response to this public concern creates an opportunity to enhance public understanding by providing accompanying information on detection. In addition, as noted by some utilities, the issues would be enhanced by partnering with or involving highly credible third parties. Trust is seen as critical to the water industry. Did the apparent surfacing of this issue by the media stigmatize the wastewater and water treatment industry by conveying to the public that the industry did not apply appropriate controls or treatment approaches? Could this be likened to the apple industry Alar scare in 1989 or the bovine spongiform encephalopathy (BSE) scare in 2000? The water industry is already in a place where risk-induced stigma is prevalent, so this could be a relatively high priority area for additional research.

It is premature to assess the effect evolving public opinion will have on future treatment process decisions, reuse, and biosolids management programs. The resultant public perception of the issue is clearly one of elevated concern over the presence of trace organic compounds/ pharmaceuticals in wastewater effluent and receiving waters; in drinking water; and potentially in all treatment residuals. This is fertile ground for anyone wishing to plant seeds of fear regarding publicly owned treatment plant discharges, water treatment processes, and reuse and biosolids programs. New policy initiatives may be needed to encourage partnerships with third party experts and organizations. Such initiatives could include communicating broadly with the industry about the means to discuss the issue of trace organic compounds/pharmaceuticals in water. This issue should be discussed in a manner designed to explore the relationship between the effective variables of terminology, images, information, and risk perception. If utilities work with the media to help with the information flow about the issue, they might be able to attenuate the risk by putting it in context.

CHAPTER 5.0

Communication Framework

5.1 Putting Activities and Ideas into Context: A General Communication Framework for Utilities

Based on the literature review, a general framework for communications was developed, as described in this section. The general communication framework was prepared to illustrate steps in the communications process and place actual communications (e.g., brochures, public meetings, tours, webinars) in the perspective of the bigger picture of activities that utilities can undertake to monitor and improve the effectiveness of communications. The general communication framework was developed by organizing communication principles, practices, and recommendations gleaned from the literature review into several key steps. The framework is meant to be useful for utilities to see their communications in the context of a bigger picture of conscious and directed activities that support their overall communication vision and goals. Utilities may be able to think through the steps reflected in the framework to improve their existing communication program on trace organic compounds. This framework can be used to convey a wide variety of messages, including those relevant to trace organic compounds as well as a wide variety of other potential health and environmental communication topics.

Several other frameworks were reviewed that summarize the elements of successful communication (see, e.g., Table 3-1 and the related description in Section 3.5.2). During the 2007 WERF workshop on trace organic compounds (described in Section 1.0), the working group which was discussing risk communication-related research plans developed a schematic illustrating a cyclic communication process. They presented this figure to the rest of the workshop participants as an introduction to how they thought about the communication process. Key aspects of these frameworks were integrated into the general communication framework shown here, including steps in the communication process and the iterative nature of communication plan refinement and adjustments.

The general communication framework emphasizes the following key steps in the communication process:

1. Understand stakeholders and issues.
2. Define communications challenge, viewpoint and key messages (i.e., develop a communications plan).
3. Align program with viewpoint (i.e., align programs with the communications plan).
4. Disseminate information (followed by monitoring results to once again better understand stakeholders).

The majority of the steps also provide opportunities for the utility to build or strengthen public trust. These steps are illustrated in Figure 5-1. Of these steps, the majority of the work is conducted "behind the scenes" to make sure that the communication style and content will be effective in achieving the organization's communication goals before communicating with the public and media. This figure calls attention to the importance of building and maintaining trust throughout the communication process and the central role that trust plays in successful communications. This figure also illustrates the iterative and adaptive nature of successful communication, as information from each step in the framework feeds into the next step of the process.

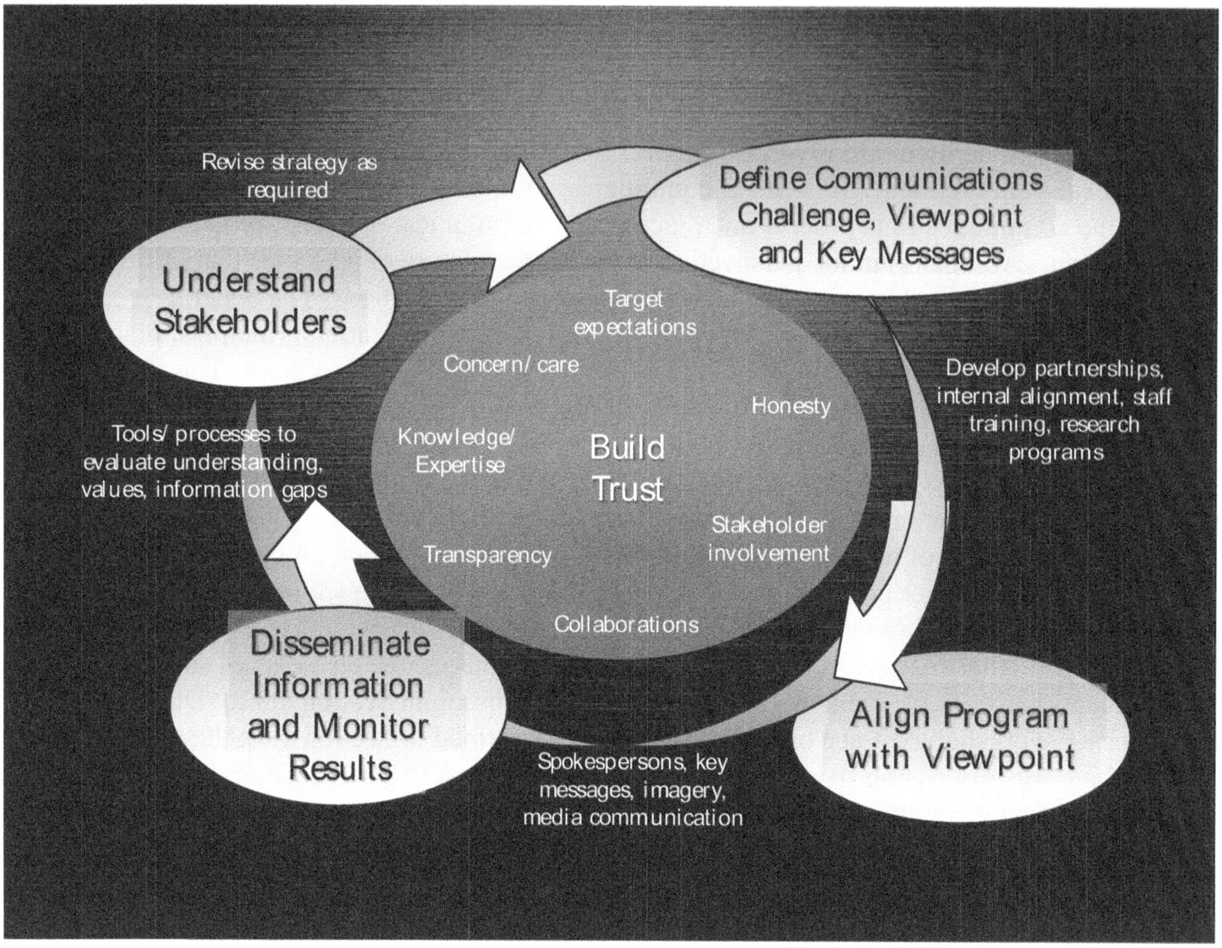

Figure 5-1. Communication Framework.

Each step in the communication process is described in more detail in the following sections.

5.1.1 Build Trust

Trust is the central facet of the communication model. Based on the literature review, increased trust will facilitate the communication process; the degradation of trust will make the process harder. Trust can be built or degraded as entities interact as a result of their decision-making processes, communication of intentions, and follow-through on plans and promises, as discussed previously in Section 3.0. Because trust is built through a process, it is not presented as a discrete step in the communication model (e.g., it is not as simple as first building trust and then communicating). Instead, trust-building is accomplished during several of the other steps in the communication process as stakeholders interact; gather information about each other; process, respond and act on that information.

Trust can be increased through efforts to *understand stakeholders and key issues*, as this work speaks to the utility's concern/care and knowledge/expertise, factors that have been shown to reinforce trust. The outcome of efforts to better understand stakeholders and key issues will permit utilities to better direct their efforts to target and meet stakeholder expectations, another factor shown to be increase trust. (Unrealistic public expectations that are identified can also be addressed in a communication program; for example, via public education campaigns.) Finally, efforts to understand stakeholders will result in greater involvement of stakeholders in the planning and decision-making processes, another way of building trust. (This form of decision-making provides an opportunity for frequent interaction with stakeholders, which provides more opportunity to develop trust, and also clearly sends a message of concern/care for different stakeholder groups and places a value on transparency in the decision-making process.)

Trust can also be increased through efforts to *align programs with the communication plan and viewpoint*. For example, a utility may decide to partner with the local public health organization or other trusted organizations and individuals. Based on the literature review, this approach will increase public trust. Decisions to participate in research programs can build trust as utilities will be improving their expertise and demonstrating leadership, concern and care.

During *information dissemination and monitoring of results*, trust can also be built. For example, changes in the utility's key messages to target key issues can increase public perception and trust, similar to the impact of a branding campaign on product reputation. Changes in outreach methods (e.g., to reach minority groups or special interest groups in the community) can increase trust by demonstrating care and concern, as well as reaching a wider audience. These principles apply when communicating with the public through the news media as well.

5.1.2 Understand Stakeholders

The importance of this step in the communication process was highlighted in the literature we reviewed and summarized in Section 3.0, including underlying psychological research findings of people's reactions to risk, as well as the different values, fears, objectives and definitions of success brought to an issue by the public and other stakeholders. People may become frightened or alarmed about trace organic compounds. Sometimes, the facts are alarming. However, people can also become unduly alarmed by misinterpreting information or jumping to conclusions. For example, it is common for people to assume that if a compound harms the reproductive cycle of an aquatic species, it will likely harm humans as well. This is an example of "intuitive toxicology," people's intuition or best guess about toxicology, though it is not necessarily supported by science. Fish and other aquatic organisms spend all of their lives in the water and use it as a supply of oxygen. Therefore, they will be exposed to higher doses than humans. In addition, hormones involved in fish reproduction and development are different than the analogous process in humans. Understanding people's concerns and thought-processes may not be <u>the</u> key to successful and easy communication, that is, people's sense of "intuitive toxicology" may persist after they are provided with facts; however, it is a first step towards improved communication and may shed light on more effective communication strategies.

Stakeholder understanding of key issues related to trace organic compounds is still largely unknown. This is highlighted in Section 6.2 as an area for future research. Recent interviews with utility representatives highlighted a number of public concerns about human health effects of trace amounts of pharmaceuticals, perhaps as a result of the AP article's focus on pharmaceuticals detected in drinking water supplies. Those interviewed indicated that the public and sometimes journalists do not have enough background to understand the context of issues that are covered by the media and may be confused. Human health effects may be an area for improving public perceptions of the risk of trace organic compounds via better communication programs. Conversely, information campaigns focused on public health and safety may merely serve to remind the public of the presence of contamination or origins of the water, creating alarm (WateReuse Foundation, 2004b).

Focus group findings in previously published studies and utility interviews have shown that the public does not have the same context as utilities for understanding water quality monitoring programs and analytical methods. The public may not understand the significance of detection limits or concentrations detected in water supplies. During interviews with utility representatives, one utility employee noted that people often have a higher expectation of technology performance and ease of monitoring. People assume that, in this day and age, monitoring is as simple as collecting a water sample and running it through an analyzer to get a result. Based on this simplified concept of monitoring, people therefore do not understand why utilities are not routinely testing water supplies for trace organic compounds, just to be on the safe side. In addition, people do not recognize the importance of detection limits. People base their impression of water quality as good or bad, healthy or unhealthy, on the presence or absence of a compound. When thinking about contaminants, people may use a "contagion model" where the presence of any amount of a substance appears to be harmful, regardless of concentration or detection limit. (Think of a small fly in a large bowl of food – the natural reaction is to get it out of there, regardless of its size!) Giving people health-based guidelines for interpreting concentrations is critical to helping them understand whether or not detections equate to risk. Utilities may hesitate before monitoring for compounds for which there are no established health-based guidelines because they will not be able to tell the public whether or not detections are a potential concern, or provide a context for interpreting the result.

Understanding the perspectives of other stakeholders, aside from the general public, is also important for successful communications. Depending on the complexity of the issue, (based on our literature review and interviews with utility teaming partners), stakeholders might include the general public, community groups or public interest groups, the media, regulators, politicians, other utilities, Native American tribes, local businesses, developers and other organizations.

Understanding media interests and motivations, format restrictions, information sources and the impacts of media coverage is a priority for utilities. Media coverage has amplified public opinion on trace organic compounds and served to spotlight the issue; public attention and actions rise and fall with news coverage. Judging from the frequency of media article coverage, concern over pharmaceuticals and other trace organic compounds has grown over the last few years. The importance of images, words and ideas expressed in the media can have a ripple effect on public concern, resonate with a number of psychological and social principles (dread, trust, involuntary exposure, dose response confusion, and inability to distinguish differences between aquatic studies and human health impact). More information on this topic is summarized in Section 3.4 and is included in the conclusions to Section 4.0.

Regular communication with regulators will help utilities stay informed about upcoming regulations and ensure that utility messages are consistent with messages from the regulatory community. In the case of trace organic compounds, utilities are often participants in the latest studies and are familiar with the practical considerations of monitoring and treatment. Their input is therefore invaluable to regulators and can help design future regulations, ensuring that regulations are informed by science and can be smoothly implemented. The perspectives and priorities of regulatory agencies on trace organic compounds vary at a regional, state and Federal level and may change with findings of scientific research, public interest/media coverage, attention from Congress, and political climate.

After understanding stakeholders' concerns, utility can ideally pinpoint differences or information gaps between their own perception and other stakeholders' perception of the same topic. These areas may be addressed, directly or indirectly, by future communications.

5.1.3 Define Communications Challenge, Viewpoint, and Key Messages

The focus of this step in the communication process is to look within the organization, increase self-knowledge, and develop messages that are consistent with the utility's objectives and values. The process is similar to developing or re-aligning a branding campaign. This is a topic that has been well-studied in advertising disciplines. Communication managers may want to start by brainstorming with other employees or with an outside party to identify three or four key words that capture the essence of the utility's values and behavior. These concepts can be later integrated into key messages.

Next, utilities can formulate statements about their philosophy, their priorities and the objectives of their communication program on trace organic compounds. This can form the background for developing key messages and will ensure that the messages are consistent with utilities' values, are believable and are easily recognizable. As part of this process, utilities can also identify images that are consistent with their values, as images are a powerful way to convey messages. More details on the benefits and process of branding were described in Section 3.5.1. Finally, the utility will need to identify several key messages to convey to the public. These will likely vary from one utility to another. Utilities with state-of-the-art treatment technologies such as reverse osmosis or ultraviolet treatment that are producing purified drinking-water quality water may focus their messages on their treatment process as an effective barrier against the presence of these compounds in water supplies. On the other hand, drinking water providers with relatively pristine source area/watershed catchment area may want to focus their communication efforts on source control activities. Drinking water providers drawing from surface waters that are heavily impacted by upstream wastewater discharges might focus their attention on collaborative regional efforts to cooperatively conduct monitoring, source control activities and treatment processes.

5.1.4 Align Program with Viewpoint

After defining the utilities' values and coming up with a "viewpoint" or communication vision that succinctly describes and conveys these values through message content, images and associations, the next step is to align the utility's program with their vision. This process is critical to building trust and message consistency, as communications that are contrary to actual operations will be perceived as hypocritical, degrading trust. It is important for a utility to support this process at all levels of the organization, which requires support from upper management. Sometimes it is helpful to bring in an outside party to conduct the review and summarize key findings anonymously with recommendations about ways to improve the organization. Other potential positive outcomes of this review process include changes in organizational programs, better communication between different departments, streamlined communication practices and cost savings from improved efficiencies.

Collaborations and partnerships with other organizations are formed at this stage of the communication process. Collaborations with credible organizations can greatly increase public trust and confidence and can leverage the available resources to address communication issues. Examples of respected professions and organizations include medical doctors, pharmacists, non-profit organizations, schools and universities, research foundations and government agencies. Partnerships do not need to be formalized. Examples of informal partnerships include quoting a reputable scientist in a press release or having a government representative available at a public meeting to answer questions.

Training is another critical component of aligning the utility's communication program with their communication vision. Utilities' process for responding to public inquiries may or may not be formalized. Some utilities have multiple spokespersons that interact with the public. Calls may be directed to one of several different technical employees who may not have any communication training. Training can help utility employees to understand the utility's vision, align practices with this vision and deliver key messages to the public in terms that they can understand. Training can also help employees to anticipate and prepare a response to questions that are frequently asked by the media in a format that journalists can easily use. During training, utilities can also discuss when to interact with the media. Some may wish to be proactive and contact reporters with press reports or offer quotes whenever an issue arises. This will help to get key messages out to the public and has a greater likelihood of generating positive press. It is important that utility spokespersons and communication managers believe that the approach that the organization is taking; else, they will not be effective spokespersons.

An example of "alignment" is where a utility decides that it values being proactive, responsible, and collaborative. Alignment might argue for funding of additional programs, participating like research studies, and education and outreach activities. Any activities that support and substantiate the utility's communication vision would be performed under this part of the communication process.

5.1.5 Disseminate Information and Monitor Results

The fourth step in the communication process is to conduct outreach and monitor the results. Several different methods for disseminating information are recommended in order to reach a variety of different people. Examples range from mailing information to people's homes with or without their utility bills to conducting water/wastewater education or treatment system tours. More details about message delivery are presented in Section 3.5.3.4.

A number of different guidelines have been published for spokespersons in order to make them more effective in reaching the public with their messages. General guidelines are summarized in Section 3.5.3.5; guidelines for high concern/low trust situations are summarized in Section 3.5.3.8. Basic principles to follow at a public meeting include demonstrating concern and care, knowledge and expertise, honesty and openness and dedication and commitment. Prior preparation and advance planning can facilitate a smooth presentation. Another piece of advice is to stay focused on key messages (typically a maximum of three), followed by information that supports these key messages. In preparing to interact with the media, key messages should be disseminated in a format that meets the needs of the media – for example, short one- to three-line quotes or sound bites for broadcast media. In low-trust, high-concern situations, the same communication principles apply. However, messages must be greatly simplified so that people's brains can process the information. When people are stressed or upset, they have difficulty processing and remembering information.

The importance of pretesting messages and polling perceptions before designing and launching an information campaign was described in Section 3.6.3. Messages can be pretested using one or more focus groups in order to gauge whether or not people will respond to communications as intended and whether the communication materials generate additional questions or concerns. Pretesting messages is a common practice before launching an information campaign. Results from this step feed back to understanding the impact of the communication program on stakeholders, reevaluating and refining key messages if necessary, realigning utility practice with the communication vision and again disseminating messages and monitoring the results.

CHAPTER 6.0

SUMMARY AND RECOMMENDATIONS

Water and wastewater utilities and other WERF subscribers have identified a need to better communicate with the public and with the media about the presence of trace organic compounds in water supplies and the associated potential risks to human health and the environment. As the first phase of research to address this need, WERF sponsored a review of the research previously published on risk communication and a summary of key findings. Results are presented in Section 3.0 of this report. An analysis of the way that the media presents information about trace organic compounds was also conducted, including the key messages, context, choice of vocabulary and imagery, the likely impact on the public, and other attributes (Chapter 4.0). Representatives from ten different utility teaming partners shared their perspectives on communicating with the public and with the media on trace organic compounds. They described their outreach programs, communication methods, key messages and thoughts about communication deficiencies or needs (Chapter 4.0). Based on the literature review, a generalized communication framework was prepared for utilities to place communications (e.g., brochures, press releases, educational tours, etc.) in the context of a mindful process of understanding stakeholders, communication goals, aligning programs and messages communicated to the public with those goals, and monitoring the impact or effectiveness of communication materials (Section 5.0). Take-home messages from this work are summarized as recommendations for utilities in Section 6.1, and as recommendations for further research in Section 6.2.

6.1 Recommendations for Utilities

A list of recommendations for utilities was compiled based on key findings from the literature review, lessons learned from risk communication in other industries, analysis of media articles, and the perspectives of ten different utility partners gathered during interviews. These recommendations are prefaced by the realization that utilities today, particularly small-sized utilities, are often busy with day-to-day activities, plans or projects to meet upcoming regulations, population growth and/or capital improvements. Utilities may not have the budget or staff time to devote additional resources to the issue of risk communication, public education and outreach, or to develop a communication program specifically addressing trace organic compounds. Research recommendations were therefore prioritized, and are listed with the understanding that efforts may need to be scaled to a particular utility and their public in accordance with the importance of trace organic compounds:

1. ***Talk with stakeholders.*** Gain a perspective on the public and other stakeholders' values, interests, and priorities; their impressions and understanding of the issue of trace organic compounds; their understanding of other water issues such as maximum contaminant levels (MCLs), water and wastewater treatment technologies; the importance of these issues to the public; their attitudes towards the utility; who they trust as sources of information on the issue; key concerns; suggestions for addressing these concerns; and

other topics. This understanding can be gained by talking with the public or conducting focus groups or surveys. Summarize this information and identify gaps in information or perception (i.e., areas where peoples' understanding of the issues or perception of the organization are not aligned with the utility's practices and values). Some of these topics may result in changes to key messages or provide the incentive for adopting new programs and practices. More details are provided in Section 3.3.

2. ***Understand the media's needs and tailor communications accordingly.*** Prepare a list of guidelines and talking points for utility spokespersons to use when approaching the media or responding to media questions. Proactively contacting the media can lead to positive press, build relationships with journalists, and allow the utility to frame the story rather than reacting to someone else's portrayal of an event. Keep in mind that the media is not interested in educating the public but instead seeks to hold the public's attention (Sections 3.4 and 4.0). Communications with broadcast media may be limited to a brief sound bite or one-to-three line quote. Press summaries can be prepared to highlight utilities' key messages and summarize who, what, when, where, how, and why. Supporting documentation can be made available on the utilities' website (Section 3.5.3.6). Messages must be further simplified when responding to media inquiries during a stressful time, as people have difficulty hearing, understanding, and remembering information when they are upset (Section 3.5.3.8). The media sometimes portrays water and wastewater utilities unfavorably. However, there is opportunity for utilities to provide the media with their facts and perspective on this topic and get positive press (Chapter 4.0).

3. ***Collaborate with other organizations***. Develop partnerships with other organizations to conduct research, develop messages, and conduct outreach as needed. For example, statements about human health effects could be developed in partnership with the local or state health department. Information about pharmaceutical take-back programs could be endorsed by or distributed by medical doctors and pharmacists. These suggestions stem from conversations with utility teaming partners, as described in Section 4.3.3. Collaborating with organizations that are trusted by the public can increase the impact of messages delivered (Section 3.2.2.5) and strengthen support for key messages.

4. ***Involve staff in the communication process.*** This step in the communication process is about "walking the talk" and making sure that organizational behavior and messaging at all levels of an organization are aligned with messages from the communications department. If an organization is aware that it is sending the public mixed messages, it is important to involve staff to remedy the problem. Because the organization's staff contributes to the face of the organization, staff needs to be involved in the process of message development and dissemination. Activities conducted during this process are both diagnostic and remedial. Utilities can conduct interviews with staff at different levels and departments to understand how the utility is viewed internally, gauge the consistency of the organizations' values and behavior, and identify areas for improvement. They may want to review current practices to determine whether any are contrary to the communication vision (i.e., "media-proof" the organization against behaviors that could make the utility feel embarrassed if juxtaposed with the organization's values in a news story). This process will strengthen the effectiveness of

key messages relayed to the public and media by the utility and help build public trust. More information on this is provided in Section 3.2.2.1.

5. ***Define the utility's communication vision.*** Identify the utility's key values and the overall objectives of the communication program, i.e., the utility's communication vision. Express these values in a few words that summarize the defining characteristics of the utility and incorporate these concepts into key messages. This recommendation is based on literature summarizing the well-established area of brand development and the benefits of branding to convey a few key messages consistently in an easily-recognizable way. When developing their communication vision, utilities must consider their own goals, objectives, and values as well as those that will resonate with the public. (See Section 3.5.1 for more details.)

6. ***Recognize that education and communication is an iterative process***. Communication methods and key messages will be continually refined in response to changes in public understanding of the issues, public response to utility messages, realignment of utility outreach and the development of new scientific information. The circular nature of the communication process was highlighted by participants in the 2007 WERF workshop and in published communication plan elements summarized in Table 3-1.

7. ***Collaboratively prepare and implement a communication plan.*** Work with other stakeholders to prepare and implement a communication plan (Sections 3.5.2, 3.5.3.2, and 3.6.5). When working with different stakeholders, ask each of them to define success and share their definitions with the group. Adopting inclusive decision-making processes provides other stakeholders with more involvement, control, and choice, reducing the perception of risk. It provides them with more opportunity to become familiar with the issues and also increases their understanding. Accepting other stakeholders as partners in the process also builds trust. (See the discussion in Sections 3.2.2.4 and 3.6.2.)

8. ***Conduct communications training.*** Provide training for all staff members who are interacting with the public or with the media, as described in Sections 3.5.2 and 3.5.3.1. According to utility teaming partners, multiple spokespersons are often involved in responding to questions from the public. Many of the utility employees fielding these questions have technical backgrounds (Section 4.3.3). Topics to cover during training include the utility's communication vision, basic communication principles, preparation for frequently asked questions from the public and the media, and a discussion of the utility's ongoing programs and collaborations in different departments on trace organic compounds.

9. ***Develop key messages.*** Develop messages that reflect the utility's values and viewpoint. There is a general lack of resources available to help utilities determine what the key messages on trace organic compounds *should be*. There is no consensus on key messages about trace organic compounds from the water industry. Due to the complexity of the topic and relative newness of research on this topic, scientific data do not support generalized assumptions. Different organizations will likely have different messages based on their communication goals and organizational purpose. However, key messages can still be communicated in the face of this uncertainty. For example, proactive efforts

such as educational programs, participation in research, pharmaceutical takeback programs were key messages described in communications with the public and media by one of the participating utilities (Chapter 4.0). As suggested in Section 6.2, further research is recommended to develop consensus messages from the water industry and communication tools and techniques for utilities in communicating those messages effectively. The analysis of media coverage of key messages (Chapter 4.0), impacts of vocabulary and imagery, and discussion of benefits are relevant to this step in the communication process (See Sections 3.5.3.3 and 3.5.3.4 for more details). For example, emphasizing key benefits has been shown in increase people's overall affective response, reducing perceived risk (Section 3.6.4). During message delivery and other communications, build trust by emphasizing concern/care, honesty, transparency, and knowledge/expertise. Train spokespersons in these skills and other communication techniques as described in Sections 3.5.3.5 and 3.6.2.

10. ***Pre-test messages and monitor results.*** Conduct focus groups or surveys to predict and confirm the impact of potential messages, choice of vocabulary, and choice of images before using brochures or other processes to reach the public. Use psychologists or cognitive scientists to inform the study design and analysis. This step can save much expense and damage to one's reputation. This is one of the key findings from the review of other industries' risk communication experiences (Section 3.6.3).

11. ***Review findings on the psychology of risk perception***. The researchers suggest that communication program managers and other interested staff at utilities read some of the social psychology literature (reviewed in Section 3.1 and 3.6.6) to become familiar with the way that the public thinks about risk, risk assessment, and trust. Social psychology research on topics such as the contagion model, "intuitive toxicology," affective reasoning, and stigma can help utility staff understand and predict the public's reactions. Uncovering common ways of thinking about the issue and how people make decisions can help scientists better understand how the public processes information and can better inform messaging. Ideally, a communications manager at the utility would have a background or training on these topics or would review key references cited in Section 3.1 of this report. Other utility staff would be exposed to these ideas during communications training.

12. ***Support a full-fledged water education program and risk communication strategy on trace organic compounds.*** Such a program would establish and fund an outreach (education) and communications program to provide clear and consistent messages to staff, the public and other stakeholders on trace organic compounds. This recommendation could be implemented after the utility identifies key messages on trace organic compounds. One of the key findings from the literature review of other industries is that the public does not understand detection and concentrations. Communications that present these issues in easily understandable context can counteract the more sensational press articles and enhance an organization's reputation by allowing them to be the source of easily understandable information (Section 3.6.1). Nine out of 10 utilities interviewed during this study felt that trace organic compounds were either very important or important to their organizations. Only one utility classified the issue as somewhat important (Table 4-4). The fact that the utilities see the topic as important reinforces the

need to help the public frame the issue appropriately. Experiences at two of the utilities demonstrated that proactive outreach to the media could yield great understanding and diminish sensational reporting.

Many utilities will be looking towards industry associations and organizations such as WERF, WEF, and AwwaRF to provide them with "short-cuts," tools, templates, and readily available, easily understandable summaries of key messages. Resources are not available for most utilities to conduct background research on public opinions within their jurisdiction, develop and implement communication programs specific to trace organic compounds, and prepare brochures and other educational materials to distribute within their community. Some utilities, including a few of the ones interviewed in this study (Chapter 4.0), recognize that trace organic compounds are a priority and are conducting outreach and supporting educational activities on this topic. However, they may be the exception. Section 6.2 summarizes ways that WERF and other organizations can assist utilities by collaboratively funding additional research and developing educational materials on trace organic compounds.

6.2 Suggested Future Research

Based on the review of the literature, media articles, and interviews with different utilities, several recommendations for future research were developed. Many of these research needs stem from the recognition that communication needs are shared by a variety of different utilities and other organizations. Interorganizational collaboration may be an effective approach for delivering consistent messages on this topic. Another common theme is that very little information specific to risk communication about trace organic compounds has been published to date. The following recommendations for further research have been placed in order, recognizing that some research projects precede others:

1. ***Conduct research and develop guidance to inform utilities' messages on trace organic compounds based on public understanding and response to a variety of issues.*** The literature review indicated a need for more research on communications specific to trace organic compounds, guided by the work conducted by social psychologists. There are challenges to communicating with the public when issues are fear-based, such as the presence of trace organic compounds in water supplies. Some of the questions that could be answered or guided by the research include the following:

    - *Detection and concentrations:* How specifically do utilities address the public's expectations of zero risk? Most importantly, how does one bring a perspective of proportion to this dialogue by discussing the issues of concentration and detection? The public may not have the same context as utilities have in understanding water quality monitoring programs and analytical methods. They may not understand the significance of detection limits or concentrations detected in water supplies, and may equate detection with harm, as described in Chapters 4.0 and 5.0. Some people don't understand the difference between parts per trillion and parts per million and tend to focus on the number reported versus the improbability of harm.

    - *Human health concerns:* A key finding from utility teaming partner interviews was that people are worried about human health effects more than ecological effects.

People believe that what happens with animals can happen to humans. It therefore may be challenging to convince people that compounds that are feminizing fish are safe for them and their children to drink. How do utilities and other practitioners overcome emotional images? Will information campaigns focused on public health and safety be reassuring, or will they merely serve to remind the public of the presence of contamination or origins of the water, creating alarm? The experience of perceived risk is not only the scientific evidence regarding harm but also the result of the media/public perception by which the public interpreted the risk.

- *Pharmaceutical take-back programs and other partial solutions:* After understanding the issue, the public may want to know what can be done to reduce risks and minimize the impact on the environment. Unfortunately, the answer is complex due to the variety of different compounds that may be present at varying concentrations, analytical methods, treatment technologies, removal efficiencies, and levels of potential concern, not all of which are known at this time. When fielding questions from the public, some utilities have emphasized steps that the public can take to minimize the problem, such as bringing unused medicines to a pharmaceutical take-back location instead of flushing pills down the drain. Utilities might wish to explore whether recommending pharmaceutical take-back programs or proper disposal programs is helpful. Questions to explore include whether such programs attenuate or amplify the perception of "risk" associated with pharmaceuticals in water. Pharmaceutical take back programs are far from a complete solution to the issue because many pharmaceutically-active compounds are not metabolized in the body and pass through the human body into human waste streams unaltered. What are the cost to benefit ratios of these programs and does this ratio compare favorably with the cost/benefit of additional treatment (dollars per kg pharmaceuticals removed)? Does the public understand that disposal of unused pharmaceuticals down the drain is just one component of the concentrations entering the environment? Does the public perceive the programs as a way to change the amount of material detected for these compounds? Does the public understand that some treatment technologies such as oxidation and reverse osmosis are able to remove trace organic compounds? Will the public be willing to pay for the increased costs of removal if there are no findings of harm to human health and the environment? How supportive is the public about alternative wastewater management strategies. Is there a high demand for monitoring? How can monitoring results be meaningfully communicated?

- *Other:* Other questions include how utilities can communicate effectively with the public about trace organic compounds, despite uncertainty. What can they do in the interim while lots of information is not known? In addition, how can utilities answer questions about trace organic compounds, including pharmaceuticals and hormones, in treatment plant effluent, make different levels of wastewater and water treatment understandable, work on a watershed basis to ensure that not all constituents in the water are defined as an end-of-pipe problem or that both wastewater treatment and drinking water is understood, and potentially involve the public in product substitutions.

2. ***Survey utilities to identify common goals and explore areas for inter-organization collaboration.*** Additional research is needed to identify common ground and develop a shared vision for utilities that are dealing with the issue of uncertainty and risk communication on trace organic compounds. Utilities likely know that they want to be viewed as "doing the right thing" and earn the public's trust. However, they may not have a sense of what consumers want to see/hear before they will trust utilities. It seems there is a need for the industry to get together and settle on three or four core messages that can be the basis of an informational campaign or risk communication strategy for the group. There is a need to come to consensus about what these messages are from an industry-wide perspective.

   Utilities likely do not have the resources to develop and implement a communications program that addresses all of the steps in the communication process in-depth. In particular, small utilities face this challenge as costs are a higher percentage of their annual operating budget. Consolidation can also be used as a way to improve message consistency to the public. Given potential benefits, utilities may want to establish an interagency collaboration/branding campaign to communicate shared messages to the public.

   To determine whether this idea is viable, a study could be performed to analyze the similarities and differences between utilities' key values, messages and priorities, explore whether or not utilities would find a national spokesperson or organization useful and identify the amount of common ground that utilities have. For example, what are the differences between communication needs and messages of publicly- and privately-owned/operated utilities? Between small, mid-sized, and large utilities? Do water and wastewater utilities share the same concerns? Often wastewater utilities are not treating wastewater to drinking water quality standards and therefore do not have the same technologies or level of protection that drinking water treatment plants or wastewater utilities with advanced treatment capabilities have. How much message content, if any, would be lost by focusing on messages shared by the industry as a whole? This study could involve a national survey and/or interviews to identify utility representatives' thoughts on communication-related needs, communication visions, overall goals, and objectives; interest in supporting a national program to develop messages and conduct outreach on their behalf; summarizing the results to identify whether or not such a program would be viable; and defining messages that are shared by a number of different utilities. This work would expand on the preliminary interviews conducted in this report and produce results that were representative of utilities nationwide.

   Alternatively, other organizations could be interviewed to determine their willingness to communicate about these issues and/or provide funding for risk communication efforts. Examples include non-profit environmental organizations, health and medical organizations, and applied research foundations. The amount of trust the public places in each type of organization could be surveyed, as well as the impact of financial contributions or compensation on this level of trust.

   Organizations such as WERF, WEF, AwwaRF, the WateReuse Foundation, NACWA, NWRI, U.S. EPA, and others have funded applied research on risk communication in the past. Funding a single group to develop consensus-based talking points, key messages and press statements may be more consistent and more efficient compared with each organization preparing its own version of the same issues. Research organizations would need to be polled to gauge their interest in collaborating on this issue and identify areas of overlap.

3. ***Develop tested communication tools and templates for utilities.*** Tools could include key messages and accurate language and words to use, sample press releases, an editorial or article for local papers or publications, a small brochure about trace organic compounds that could be printed by an agency, a "how to" guide for responding and seeking out the media, an in-house training guide for non-communications staff, and/or a list of links for specific articles and research. A brief (five-minute) video about trace organic compounds could be made available for use by utilities on their websites, as part of a speakers' bureau, and for other presentations. Utility partners and project reviewers have suggested putting together a list of simple message points and communication tips for dealing with the public and the media (brief "cheat-sheets") and preparing materials that could put the issues of detection, concentration, and other scientific concepts into perspective for the public and the media. Another suggestion is to test the effectiveness of existing communications materials prepared to date by organizations such as WERF's online Q&A on pharmaceuticals in water (WERF, 2008b) and WEF's online Resource Paper for Public and Press, *Microconstituents in the Water Environment* (WEF, 2007a).

4. ***Conduct a detailed analysis of the impacts of media coverage and develop recommendations for utilities to interact with the media.*** As illustrated by the March 2008 AP coverage of pharmaceuticals in water supplies, media coverage of an event can amplify and direct public attention. Research could be conducted to expand on the preliminary media analysis presented in this report (Chapter 4.0) and develop guidance for utilities in communicating with the media about these issues. Understanding media interests and motivations, format restrictions, information sources, and the impacts of media coverage is a priority for utilities so that they can better convey their messages to the media. Media coverage has amplified public opinion on trace organic compounds and served to spotlight the issue; public attention and actions rise and fall with news coverage. Judging from the frequency of media article coverage, concern over pharmaceuticals and other trace organic compounds has grown over the last few years. The importance of images, words, and ideas expressed in the media can have a ripple effect on public concern; resonate with a number of psychological and social principles (dread, trust, involuntary exposure, dose response confusion, and inability to distinguish differences between aquatic studies and human health impact). More information on this topic is summarized in Section 3.4 and is included in the conclusions to Chapter 4.0. Research could be conducted to answer questions including the following:

   - Does media coverage amplify which water issues receive attention to the point of defining public priorities? For example, during the AP coverage of pharmaceuticals, the water and wastewater industry was mounting a campaign to focus on the need for more infrastructure investment. How does the public prioritize water investment issues?

- How do journalists address the issue of trace organic compounds in their investigative reports? Do journalists perceive that there is a major health issue that needs to be exposed? Do journalists consider the water industry to be a partner in telling the story, as an adversary or other? Results from this research might help utilities to better interact with the media, understand media pressures and needs and opportunities to get positive press.

- Which media articles, headlines, cartoons, and artwork create the most alarm? Is impact linked to how memorable these statements and images are? Can memorability be characterized as a function of affect conveyed and content (i.e., providing fish with human characteristics)?

- What role does stigma play in public communication on trace organic compounds and how can utilities address it? Has media coverage stigmatized the wastewater and water treatment industry, similar to the Alar scare in 1989 or the BSE scare in 2000?

- When and within what context should utilities communicate with the public about trace organic compounds through the media? Based on our interviews with ten different utility partners, many of them wish to be proactive about this issue but are concerned about presenting information to the public without context, for fear of increasing public alarm unnecessarily.

5. ***Establish an online education/media center for journalists and the general public.*** There is a need to provide educational materials to the public and to the popular media that are easily understandable and accurately reflect scientific information about trace organic compounds. Such materials would provide context for the issues described, explain the issue of detection, regulatory processes, risk management philosophies, ongoing and planned research topics, treatment technologies, and other related information. Assembling this material would counter the shallow coverage that these topics are given in the popular press and provide the public with more information to inform their opinions and decisions about water resources and contamination. A website would be an appropriate platform for this effort. To make the content more appealing to a broader audience, website text could be accompanied by images, webcasts, and/or podcasts.

   The sponsoring organization could be structured as a non-profit or not-for-profit organization and employ science writers and other journalists who work closely with different utilities and other technical people. The organization could play a similar role for the press as the non-profit Environmental Working Group (EWG), an organization that is typically quoted in the popular press and whose coverage of environmental issues steers media attention and subsequent public interest (www.ewg.org). It could become a go-to source of information that utilities could reference to counter alarmist media coverage and provide more balanced framing of the issue.

   Such a media resource guide could provide easy access to research findings and motivate the media to utilize this common website for sources of information. Partners could also implement webinars for media professionals to distribute accurate information and discuss the issues that utilities face.

6. ***Spearhead collaboration between environmental organizations.*** Assuming that sufficient overlap will be identified between utilities and other organizations, WERF or other research organizations could spearhead collaboration between different environmental organizations. There have been many discussions and successful collaborations/partnerships between environmental organizations over the years. This project would involve a nationwide effort to involve large, well-known, groups whose goals and mission are similar to water-related organizations, and use the collaboration to develop and disseminate consistent messages to the public and media about water-related issues. Messages that promote clean water and a healthy environment benefit a variety of environmental groups as well as utilities. Partnering with well-founded environmental organizations would create more trust and credibility from the public and media. Such partnerships would also help educate the public and media, focus on common values, and promote more holistic understanding of key issues.

7. ***Establish centralized point-of-contact(s) and information repository for utilities.*** If found to be viable, utilities and other organizations could collaborate to hire a national spokesperson; prepare press releases and quotes for the press; list names, contact information, and qualifications of leading researchers, regulators, and utility representatives who are willing to discuss their work with the press; develop a national marketing campaign; sponsor research or expert panel review of different topics; host webcasts and/or a central website as a clearinghouse so that utilities can exchange information about trace organic compounds; post a list of references and additional resources on trace organic compounds so that utility employees know where to find the latest information and are knowledgeable about the state of science, regulations, ongoing utility programs, and/or monitoring results; post news updates and otherwise share information.

   For example, a website could be created to house collective research, papers, and reports from contributing partners (WERF, WateReuse Foundation, and others). This website would provide tremendous value and benefit to utilities, communication professionals, the media and even the general public to build understanding of communication principles and practices, as well as critical issues. The website could be promoted on all the partners' own websites as the "go to" place for information and would hopefully generate a high level of good will, objectivity, and trust with other participating organizations. A website that was easily accessible to all agencies, regardless of subscriber status, would speak to the concepts discussed in this report (i.e., transparency, honesty, and trust) and could provide a democratic and inclusive method for educating people about the key issues.

8. ***Host seminars.*** Short courses and other knowledge transfer initiatives could be conducted to help utilities and others to assimilate key concepts about risk communication and spread awareness of existing tools and resources. WERF and its partners could fund and conduct two-day seminars to describe and outline the research and recommendations from this report. Communications professionals and other technical/management staff

could attend these seminars to gain a focused understanding of the concepts and recommendations. These seminars could be held in four or five places throughout the U.S. so that the dates and locations could allow for cost-effective and time-efficient travel and attendance by a variety of utility representatives, journalists, and other interested parties. Seminars could be repeated annually and evolve as more research is conducted and other studies are completed. To encourage the participation of small agencies with limited budgets, registration fees could be waived or kept to a minimum.

9. ***Host webinars.*** WERF and its partners could also implement webinars that to tackle mini-topics from this report. As an example, a webinar could focus on media relations, or be used to train additional utility workers who may not be able to attend a two-day workshop. It would be very helpful to have a focused webinar on how people perceive and react to risk as related to water issues. Communication professionals can talk about the importance of building trust with the public and other stakeholders, but if management doesn't understand the relevance and reasoning behind its importance, then all the communication plans in the world won't help an agency change its way of working with its constituencies. Webinars could occur monthly or bi-monthly and handle a different element of the findings and research and help communication professionals and utility staff in dealing with day to day issues.

10. ***Develop regulatory guidance for utilities.*** There is also a need to compile information about regulatory aspects of trace organic compounds. One idea is to conduct a national survey and/or interviews of regulators from a variety of different international, federal, state and local agencies including participants in the contaminant candidate list (CCL) process, U.S. EPA Endocrine Disrupting Compounds Screening Program, REACH regulations, state departments and others to explain the goals of each regulatory agency, understand current and potential future regulations for trace organic compounds, approaches that different cities and states are using to address these compounds in the absence of Federal regulations, existing and potential future communications programs supported by regulatory agencies and regulators' impressions of future research needs and directions, including communication-related research needs.

    This document would be helpful for utilities as it would summarize the current and potential future regulatory developments as well as the philosophies, priorities, and perspectives of different regulatory agencies on this issue. The perspectives and priorities of regulatory agencies on trace organic compounds vary at a regional, state, and Federal level and may change with findings of scientific research, public interest/media coverage, attention from Congress, and political climate. Understanding the perspective of regulatory agencies is important for successful communications. Regular communication with regulators will help utilities stay informed about upcoming regulations and ensure that utility messages are consistent with messages from the regulatory community. In the case of trace organic compounds, utilities are often participants in the latest studies and are familiar with the practical considerations of monitoring and treatment. Their input is therefore invaluable to regulators and can help design future regulations, ensuring that regulations are informed by science and can be smoothly implemented.

APPENDIX A

Utility Interviews

Interviews were conducted with 10 participating water and wastewater utilities as described in Section 4.0. Copies of the interview transcripts are provided in this section. To maintain confidentiality, utility and participant names have been removed. The agencies are referred to as Utilities 1 through 10. Key information from these interviews was summarized and presented in Section 4.0.

Interviews with Utility/Water District Teaming Partners
Utility 1

Date
July 21, 2008

Background
Interviews were conducted with the Water Quality and Environmental Services Coordinator and the Communications Coordinator for the Planning and Public Works Department. Both individuals have been helping with media coordination and outreach. Utility 1 has a website with a one-page fact sheet on their website and has prepared a document of key messages for Utility 1's Manager. Utility 1 has been mentioned frequently by the media; however, Utility 1 hasn't kept copies of all these articles.

Questions and Responses
1. How important of a topic are trace organic compounds within your organization?

> Very important. Utility 1 has been involved for about five years since research was initiated by a local university and USGS. The effort with the university continues today. Utility 1 believes that it is important for them to continue to find out about pharmaceuticals to ensure more information for future decision-making. Utility 1's involvement has been mostly in-kind. The study was initiated in 2002 or 2003 and results started to reach the public in 2004.

2. Does your organization have a designated spokesperson to answer questions from the public about trace organic compounds? Also, does your organization have a designated spokesperson to answer questions from the media about trace organic compounds?

> The way Utility 1 handles media calls is to get staff experts to answer media questions. Typically the media calls come though the Communications Coordinator and she gets subject matter experts involved. The Water Quality and Environmental Services Coordinator is a designed spokesperson.

3. Has your organization prepared any outreach materials on the issue of trace organic compounds? If so, please describe these materials and how they were distributed.

> Utility 1 has prepared a website and talking points (see attached). They are planning our first pharmaceutical take-back program day for October 2008, in partnership with Partners for a Clean Environment (PACE) and a Community Hospital. Utility 1 is in the process of developing a longer-term take-back program that would include several designated take-back days. The October take-back day will cover Utility 1 but will accept pharmaceutical from residents within the county.

4. Have you ever used Water Environment Federation (WEF) or Water Environment Research Foundation's (WERF's) resources on trace organic compounds or microconstituents, such as their website, Technical Practice Updates or Question & Answers? If so, was the information helpful? What other resources have been useful?

> Yes, Utility 1 has used WEF resources pretty extensively as well as WERF publications. WERF has some good reports on fate and transport through different wastewater treatment processes. Three or four were applicable to Utility 1 but they can't recall the actual titles.

5. What, if any, are the principal concerns in your community related to trace organic compounds (e.g., human health concerns, environmental health concerns, other)? Please estimate the percentage of public concern associated with each.

> Public concern was not specifically quantified but drinking water concerns seem to be the primary focus of people's concerns; for example, Utility 1 has gotten questions about home water purification systems. The area is very environmentally active and the impacts of fish in the local creek are the focus of the USGS research. How these findings could translate to human health was a concern by citizens. The city has very little upstream impacts to the drinking water watersheds (high alpine and trans-basin western slope watershed), and has not conducted any sampling yet. There are many municipal water users downstream of Utility 1 but no one directly below Utility 1.

6. Has your organization developed any partnerships to assist in communicating the risks associated with trace organic compounds (e.g., partnerships with local research organizations, medical doctors, consultants)?

> Utility 1 has not directly collaborated with any one on risk communication. However they are starting to pursue some programs that will probably relate to risk communication. The Community Hospital will be the host for the pharmaceutical take-back day so, to that extent, they are a partner along with PACE. There isn't definitive information on the risks yet. A lot of this information is still emerging.

7. What tools, processes and approaches has your organization used to interact with the public? Describe the situations for which each approach has been used. Do you think that these methods attenuate public concern about trace organic compounds?

> Utility 1 has used their website and talking points. They have distributed a utility bill insert (one paragraph) to provide information on the take-back program. If there is a concern, people may call the main Public Works number, which is in the Communications Coordinator's group and be transferred to her.

8. In your geographic area, have there been any changes in practices or programs to address trace organic compounds, such as monitoring programs, treatment studies or pharmaceutical take-back and disposal programs? Were any of these changes a direct result of public concern? Were any initiated by or supported by your organization?

Regarding pharmaceutical take-back programs, Utility 1 is aware of a nearby utility that participated in a take-back program coordinated with PACE. Utility 1 has now approached PACE to conduct a take-back day in the city. Utility 1 wanted to focus efforts at the source, as this is one of the simple things to reduce items going down the drain that could harm the environment. The take-back program will not completely solve the problem but it does help, and it is something that the public can do to help. FAQs will be available at the event and Utility 1 will contact local media.

9. Has your organization witnessed any changes in public involvement or perception regarding trace organics over time as a function of setting (e.g., during a drought, in response to an event, in response to the establishment of a program, etc.)? Please describe.

There doesn't seem to be any overall change in interest with setting except that calls and interest seems to be driven by the media; there are changes in the volume of public response to media articles. The public doesn't seem to understand the issue yet. People are often confused and the issue is not the easiest thing to explain in a sound bite. It is getting easier for reporters to follow since Utility 1 is helping to educate them. People are making the link that there is something they can do to help. The Water Resources Advisory Board is involved -- they have asked about drinking water monitoring and whether there will be a pharmaceutical take-back program. On the wastewater side, Utility 1 will continue to do monitoring with USGS and the university. Utility 1 is considering monitoring on the drinking water side. USGS has done some monitoring in low impact and pristine areas and detected compounds like caffeine in some locations, not just downstream of treatment plants. The issue of detection is not really well understood. The technology has been changing. We now have the technology to measure compounds at low levels. The part that is missing is what is the threshold for concern? This is still not well understood.

10. Please identify a local media report (article, radio or TV report) as an example of how the media presents information on this issue and how your organization is represented.

Utility 1 will research this. They have not saved all the articles but have saved some of the weird ones. Utility 1 remembers an MSN/NBC report on fish that covered some of the studies going on in the area. Initially there was some concern that this was only happening locally. There was a March 2008 article in the Daily Camera about the Associated Press investigation. The Communications Coordinator drafted some talking points in response (dated March 11, 2008).

11. Has your organization identified any communication deficiencies or needs (e.g., message content, style, trust, choice of media, etc.) that would improve the water industry's communication with the media and with the greater public on trace organic compounds/ microconstituents? Similarly, what approaches have been effective? Why?

Education has been effective and is still needed. Education has helped the public understand that this isn't just a local issue -- and that the topic is an ongoing study. People may not understand the topic or have enough relevant knowledge to put it into context. There needs to be more public service education and information. This information would ideally come from a national program, not just from local utilities. Information published by an organization like WERF or WEF may have more weight. Local government may not have everyone's trust; it may be easier to believe what's coming from a research organization. Information on this topic is being used by special interests. For example, Catholic press articles are using drugs in water as a reason to say that environmentalists should join forces with religious groups to ban birth control. "Birth Control Pills Poison Everyone", National Catholic Register, used Utility 1 as an example. Utility 1 tends to be referenced frequently because a fair amount of the research has been done there.

Other Comments

Linda asked what terminology Utility 1 is using. "Pharmaceuticals", "emerging contaminants" and specifically "endocrine disrupting compounds" (EDCs) are terms that Utility 1 has used. The term "EDCs"was not used specifically for the public, as this might not mean much to the public. They do not use one standard term. "Emerging contaminants" is an umbrella term used in the state. Utility 1 noted that a consistent term would be useful.

Interviews with Utility/Water District Teaming Partners
Utility 2

Date
July 28, 2008

Background
Interviews were conducted with a lead project manager at Utility 2, media relations representative from Utility 2's Department of Public Affairs, a lead writer in this same department and the director of water supply.

Rula Deeb provided a brief overview of the project, including renegotiation of a shortened scope of work, four tasks in the project and role of utility/water district teaming partners. Rula and Elisabeth will send a copy of the teleconference notes to the utilities for review. Actual utility names will be kept confidential in the final report. A representative from Utility 2 mentioned that Utility 2 has participated in a technical literature review already and offered to send this to Rula. The focus of this scope of work is a review of literature published on public relations and perception.

Questions and Responses
1. How important of a topic are trace organic compounds within your organization?

> Note: Utility 2 uses the term "microconstituents" instead of "trace organic compounds". Utility 2 believes that the issue of microconstituents is really important but they recognize that the water quality industry is still in the beginning stages of understanding these compounds and their impact on the environment. Most of Utility 2's involvement with microconstituents has been with other agencies. There has been very little public interaction. In the state, there are several water management districts. Wastewater management is not one of their responsibilities. Outside of reuse, wastewater is disposed of by deep well injection or discharge to the ocean. Wastewater issues are primarily handled by the Department of Environment Protection. From a public standpoint, Utility 2 hasn't fielded even one call from the public asking about microconstituents. The State handles this issue and if Utility 2 did receive a call on microconstituents, they would refer the caller to the State.

2. Does your organization have a designated spokesperson to answer questions from the public about trace organic compounds? Also, does your organization have a designated spokesperson to answer questions from the media about trace organic compounds?

Utility 2 does not have anyone specifically designated to handle this topic. However, they do have staff to answer press questions. They are very active in other areas with respect to microconstituents. Utility 2 staff have been tracking the issue from a water quality perspective but are doing so proactively as there has been no regulatory requirement.

3. Has your organization prepared any outreach materials on the issue of trace organic compounds? If so, please describe these materials and how they were distributed.

No, Utility 2 has not prepared any press releases or other materials on this topic. Utility 2 is not aware of any such materials that the State has prepared on this topic.

4. Have you ever used Water Environment Federation (WEF) or Water Environment Research Foundation's (WERF's) resources on trace organic compounds or microconstituents, such as their website, Technical Practice Updates or Question & Answers? If so, was the information helpful? What other resources have been useful?

Utility 2 has used WEF and WERF resources and checks their websites often. They found the website helpful for finding information about various compounds and cost-effective technology for treatment and removal.

5. What, if any, are the principal concerns in your community related to trace organic compounds (e.g., human health concerns, environmental health concerns, other)? Please estimate the percentage of public concern associated with each.

Because there is not much public interest evident at present, it would be speculative to give percentages. From an agency perspective, the majority of public concern is environmental. For example, if you put reclaimed water into a surface canal, the public may be concerned about the environmental impact to fish and wildlife. In some areas, the main issue is environmental. There hasn't been any proof that these compounds impact humans. Deep well injection tends to diminish public concerns of all types.

6. Has your organization developed any partnerships to assist in communicating the risks associated with trace organic compounds (e.g., partnerships with local research organizations, medical doctors, consultants)?

No, Utility 2 has not participated in any risk communication partnerships.

7. What tools, processes and approaches has your organization used to interact with the public? Describe the situations for which each approach has been used. Do you think that these methods attenuate public concern about trace organic compounds?

This really hasn't been something where people have asked for information. It hasn't really hit as a public issue yet. Utility 2 has tackled a lot of other issues and has a comprehensive media program, has a pretty active website and has issued press releases on other issues. The issue of microconstituents in water is not addressed on the website.

8. In your geographic area, have there been any changes in practices or programs to address trace organic compounds, such as monitoring programs, treatment studies or pharmaceutical take-back and disposal programs? Were any of these changes a direct result of public concern? Were any initiated by or supported by your organization?

No, Utility 2 does not know of any programs of this nature.

9. Has your organization witnessed any changes in public involvement or perception regarding trace organics over time as a function of setting (e.g., during a drought, in response to an event, in response to the establishment of a program, etc.)? Please describe.

Again, the issue hasn't come up.

10. Please identify a local media report (article, radio or TV report) as an example of how the media presents information on this issue and how your organization is represented.

Utility 2 searched media articles over the past several months and could not find information related to this issue that was related to Utility 2. Utility 2 is not aware of any local articles.

11. Has your organization identified any communication deficiencies or needs (e.g., message content, style, trust, choice of media, etc.) that would improve the water industry's communication with the media and with the greater public on trace organic compounds/ microconstituents? Similarly, what approaches have been effective? Why?

The issue hasn't come up so there is no strategy needed at this time. In two or three years, more communication may be needed but not in the near future.

Interviews with Utility/Water District Teaming Partners
Utility 3

Date
July 30, 2008

Background
Rula Deeb provided a summary of the project since the initial proposal. She noted that the scope was reframed and is now fairly limited to a focused literature review, a literature review from other industries, media analysis and utility interviews, and a draft and final report. The draft and final report will protect the confidentiality of the interviewed utilities. Utility 3 will be provided an opportunity to review the interview notes and also comment on the draft report.

Questions and Responses

1. How important of a topic are trace organic compounds within your organization?

> Note: Utility 3 is not using “trace organic compounds” in communications about this issue. They use various descriptors including “pollutants”.
>
> The issue of pollutants in general is very important to Utility 3 in trying to anticipate future regulations and implement aggressive source control/pretreatment programs for permitted dischargers. Utility 3 has a very active Household Hazardous Waste collection program to keep microconstituents and other pollutants (such as mercury) out of water ways, storm waterways or anything that goes down the drain. Examples include chemicals, triglycerides and other fats and oils. Yes, the issue of trace organics is important to Utility 3 as a pollution prevention program.
>
> Utility 3 is in the process of trying to put together a pharmaceutical take-back program – the Sheriff is supportive and the program is now being evaluated by the County Counsel who is looking into legal issues.

2. Does your organization have a designated spokesperson to answer questions from the public about trace organic compounds? Also, does your organization have a designated spokesperson to answer questions from the media about trace organic compounds?

> Yes, primarily the representative interviewed, and certainly people in the engineering section of Utility 3 that handle Environmental Services. Utility 3 representative would use these people as references in answering questions from the public or the media.

3. Has your organization prepared any outreach materials on the issue of trace organic compounds? If so, please describe these materials and how they were distributed.

> Utility 3's newsletter, sent three times per year, included a general disposal guide (not focused on trace organics). This was mailed to 140,000 people. Utility 3 received about 50 calls back asking specific questions about how and where to dispose of certain items. Utility 3 is quite aware of the pollutants that can get into the wastewater stream – the Household Hazardous Waste Facility is collecting about 2 million pounds of waste a year. Some hardware stores take back compact fluorescent lights (CFLs) and batteries because of mercury content. Mercury is a major concern in the area, has been in the media more than PPCPs and is better understood. Mercury was used in the 1850s to extract gold and silver from mining ore. Residual and legacy mercury is still leaching into the waters and into receiving waters today. So the public in this region is aware of the issue of source control of pollutants in the watershed. There are fish advisories because of mercury in receiving waters. The public may think of pharmaceuticals in this context.
>
> Utility 3 has also talked about triclosan in soaps in multiple articles in their newspaper. Utility 3 will scan and send these to the project team.

4. Have you ever used Water Environment Federation (WEF) or Water Environment Research Foundation's (WERF's) resources on trace organic compounds or microconstituents, such as their website, Technical Practice Updates or Question & Answers? If so, was the information helpful? What other resources have been useful?

> When newsletters are published, Utility 3 looks at a variety of sources. The Utility 3 representative does not have an immediate recollection of looking at the WEF or WERF websites but would likely do so as the need arises and would also check other websites such as EPA, State Board, and coastal groups.

5. What, if any, are the principal concerns in your community related to trace organic compounds (e.g., human health concerns, environmental health concerns, other)? Please estimate the percentage of public concern associated with each.

> Utility 3 has received fewer than five media inquiries in the past year on this general topic. Utility 3 received two or three calls following the Associated Press (AP) story on pharmaceuticals in drinking water and several calls from citizens. Public questions focused on what Utility 3 is doing to remove pharmaceuticals from wastewater and where to dispose of their pharmaceuticals. The drug issue pops up about once a quarter using related to something that national media has brought up. An article in the regional media -- about three months ago – provided information about places you can dispose of pharmaceuticals in the area. One pharmacy is taking pharmaceuticals; Costco is taking back drugs purchased from their pharmacy only.
>
> Utility 3 is trying to keep receiving waters healthy and has looked at the use of triglycerides and triclosan in antibacterial soaps and 10 years ago started a movement to reduce fertilizers that can runoff to receiving waters. Pharmaceuticals remain an issue and they get calls from the public once in a while. Utility 3 has told the public that there is really no treatment process for these compounds without going to microfiltration or other drinking water quality treatment systems, which is not typically part of the wastewater treatment process.
>
> Are people equating the issue of PPCPs and mercury in fish? People do not equate the issue personal care products to fish health. There have been some articles about trans-gendered fish. People are warned about eating fish from receiving waters because of the presence of mercury.

6. Has your organization developed any partnerships to assist in communicating the risks associated with trace organic compounds (e.g., partnerships with local research organizations, medical doctors, consultants)?

> Utility 3 partnered with the state and local county dental societies and local dentists (300 practices in the Utility 3 service area) regarding dental amalgam separators. As early as 2000, studies revealed that the majority of mercury in wastewater was from dental operations. They may have partnered with medical doctors as well. Utility 3 source control people worked very closely with the dental society when putting together the amalgam separator requirements which resulted in mandated regulations. The county now has 99% compliance with the amalgam separator regulations and has seen a 70% drop in mercury loading (from 34 to 10 pounds). Mercury is the major pollutant that Utility 3 focuses on removing because of the sensitivity of discharging to receiving waters.

7. What tools, processes and approaches has your organization used to interact with the public? Describe the situations for which each approach has been used. Do you think that these methods attenuate public concern about trace organic compounds?

See responses to questions 3 and 5. Utility 3's Household Hazardous Waste Collection Facility – 11 years old in two months – has the goal of keeping fertilizers, pesticides, oil, batteries, etc. out of environment and water bodies.

8. In your geographic area, have there been any changes in practices or programs to address trace organic compounds, such as monitoring programs, treatment studies or pharmaceutical take-back and disposal programs? Were any of these changes a direct result of public concern? Were any initiated by or supported by your organization?

Utility 3 has talked about a pharmaceutical take-back program. The county had such a program three or four years ago but it was short-lived. One of the major issues is the Drug Enforcement Agency's (DEA's) position that any controlled substance can only be held by only a law enforcement officer or the person for whom the drug was prescribed. The Sheriff is involved and the expectation is that all the drugs collected would be incinerated. Utility 3 is aware of the fact that pharmaceuticals are coming into the plant (whatever they may be) not primarily because people are flushing unused drugs down the drain --- but rather through urine and fecal matter.

9. Has your organization witnessed any changes in public involvement or perception regarding trace organics over time as a function of setting (e.g., during a drought, in response to an event, in response to the establishment of a program, etc.)? Please describe.

The public awareness of late is certainly because of the publicity generated by the media articles (like the AP article) and occasional stories run on national TV or magazines. Years ago there was a story about young boys in Puerto Rico developing breasts because of growth hormones in chickens and this generated a public response. The interviewee wasn't with Utility 3 at the time. Not very many people realize that water isn't pure – that in all likelihood all the water has been used before and passed through the treatment plant. There is not a general awareness of these issues which tends to allow sensationalism – like Toilet to Tap – and other terminology that once killed a recycled water program. Public attention tends to be short-lived.

10. Please identify a local media report (article, radio or TV report) as an example of how the media presents information on this issue and how your organization is represented.

After the AP article, there were one or possibly two articles in the local newspaper, The Daily, about pharmaceuticals. The local water district responded and sent out a press release and an article in their quarterly newsletter describing how drinking water is filtered. A nearby water agency which handles drinking water and wastewater, also issued a press release on the purity of the water. Drinking water was the concern in the AP article but this does not relate to Utility 3. Utility 3 did not follow up specifically but did get a couple of calls. Utility 3 has told the public that there is no treatment process for wastewater agencies to remove all pharmaceuticals.
Utility 3 practices reuse for irrigation and industrial uses. Other TV and radio news coverage would have come out of SF or national news like CNN.

11. Has your organization identified any communication deficiencies or needs (e.g., message content, style, trust, choice of media, etc.) that would improve the water industry's communication with the media and with the greater public on trace organic compounds/ microconstituents? Similarly, what approaches have been effective? Why?

Utility 3 has a good relationship with the local media. Deficiencies in accuracy are on the part of the media. We are seeing the number of newspapers and local reporters shrink and more reliance on wire stories rather than local news. There is less investigative reporting. Stories are edited to fill space. Utility 3 representative cannot recall seeing any article on this topic longer than one page. There has been limited scientific coverage.

The water industry in general and Utility 3 are concerned about minimizing pollutants. Utility 3 treats about 45 mgd – highly regulated for many parameters but there is no regulation about pharmaceuticals in wastewater. Utility 3 will try to anticipate where things are going related to regulations. There is no single test for pharmaceuticals. They may do tests from time to time but are not looking at regular monitoring as there are so many different things that could be tested for that monitoring is difficult.

The water district asked Utility 3 for a special permit to discharge water highly saturated with pharmaceuticals from a test they were doing to figure out filtration efficacy. Utility 3 looked at the effect on the wastewater treatment plant carefully before giving the water district a special permit to do this.

Interviews with Utility/Water District Teaming Partners
Utility 4

Date
July 31, 2008

Background
Elisabeth provided a brief overview of the project, emphasizing that the focus is to examine the public perception of trace organic compounds. Elisabeth discussed the narrowed scope of work from the project that was originally supported by Utility 4. It was explained that WERF wanted first to obtain focused literature review on the knowledge and practice of public perception and risk communication generally, a literature review from other industries, and then obtain an analysis media articles along with the input of utility partners. Elisabeth noted that the interviews are only one means to obtain utility partners' input. Another means will be for the utilities to review and comment on the draft report. Elisabeth also noted that the summary notes from this interview will be transmitted to back to Utility 4 for review and comment.

During the course of the interview, Utility 4 provided with additional background references, including an article on recycled water, input from the public on a recent Christian Science Monitor article about a recycled water project, and a link to No Drugs Down the Drain campaign website.

Questions and Responses

1. How important of a topic are trace organic compounds within your organization?

The question started with a discussion of the terminology that Utility 4 uses when discussing the issue. Utility 4 noted that it was difficult to pin down one term since there are lots of different types of compounds out there. Sometimes the issue at hand is about pharmaceuticals alone, pharmaceuticals and personal care products, or other. The problem with using the term organic compounds or microconstituents is that the public may not understand either term, and the media may not either. Utility 4 concluded that you really have to adapt the term you use to what the media is using or feels most comfortable with. There is a limited understanding by the many new media reporters who are just starting to learn about this topic. There was an excellent science writer at the local newspaper who no longer works there. Many newspapers are making the decision to cut staff and some of the more knowledgeable writers with relevant background and experience are frequently the first to be let go because they have higher salaries that those just out of journalism school. Utility 4 would, however, be quick to correct a reporter who uses the term "micropollutants" with information that these materials are not in polluting quantities (quantities that would have a negative deleterious effect) and that almost anything could be seen as a pollutant at some concentration. Utility 4 added that there are no studies indicating human health effects from low concentrations of pharmaceuticals, though there are some aquatic effects at low concentrations.

Utility 4 and other local ocean discharging Publicly Owned Treatment Works (POTWs) have been involved with some studies with a local coastal research organization that reported minor aquatic effects in the area of the outfalls of Utility 4, and two neighboring POTWs in 2005. These agencies and several universities have since funded a comprehensive study through that coastal research organization to better confirm any potential impacts to local flatfish and identify potential chemicals and sources for such effects. Although there are some signs of EDC exposure in these fish, no clear impacts on reproduction, health, or population levels are evident at this time. One thing that Utility 4 believes important to keep in mind is that legacy pollutants such as DDT can cause endocrine disruption effects. Such effects are not necessarily caused by anything present in current discharges. Communication about these effects being associated with pharmaceuticals may be incorrect. The public feels there is some new threat out there, which might not be true. It is difficult to get across the legacy compound issue from such sources as DDT and PCBs that were banned years ago. The situation is exacerbated by the fact that the new reporters don't have a good background on these issues, so all the information is new to them.

2. Does your organization have a designated spokesperson to answer questions from the public about trace organic compounds? Also, does your organization have a designated spokesperson to answer questions from the media about trace organic compounds?

No – Utility 4 is absolutely blessed in having many people with subject area depth and such a wealth of information that any one of a team of them can answer questions from the public or the media. All of the team is trained as scientists or engineers. A person may answer questions to a point and then turn it over to another person on the team, depending on the type of question. Utility 4's Information Officer may get more questions from the public because he is usually interacting with them at earth day, fairs, etc. In general, random public inquiries are routed to Utility 4's Information Officer, who then routes them to the appropriate person with in-depth knowledge on the subject. The agency works as a team to answer questions. Reporters often directly call those that they have established a good working relationship with.

There was little interest generated by the recent Associated Press (AP) coverage. In this city, almost everyone reads the local paper, which is a major newspaper – TV coverage follows the paper. So public interest is driven more by the paper's coverage than AP coverage. More interest was generated by the January 3, 2006 news story on prescription drugs in groundwater. The TV crews came out. Prior to that, there was some information in 2003 on potential impacts on flat fish, also reported by a science writer. This story got picked up and transmitted throughout the world (China). This got a lot of press. The original news story was a good story, i.e., balanced and without a lot of inaccuracies. The science writer is a respected environmental reporter and is able to take what people tell her and present it fairly, without a strong bias. However, when other papers pick up a story it often gets distorted.

A recent press conference following presentation of the flatfish study results didn't cause as big a splash as another paper in the session by a Canadian researcher who dosed lakes with endocrine disrupting compounds got all the attention. The Canadian research was also covered in trade journals. Utility 4 gets clippings from 30 or 50 newspapers per day. A recent local news article about illegal drugs in water didn't generate any interest at all. In 2006, a TV crew came out and talked about concentrations in terms of tablespoons in the Mississippi River. A representative from the public health agency and one from the water replenishment district were interviewed. There have also been some radio ads about the issue, focused on selling home drinking water treatment units.

3. Has your organization prepared any outreach materials on the issue of trace organic compounds? If so, please describe these materials and how they were distributed.

> Utility 4 has made many efforts to be proactive on the issue of pharmaceuticals in water. Since the 2002-2003 timeframe, they have been investigating this issue and implementing source control measures. Utility 4 has had a program in place since mid-2005 for educating people about how to dispose of unwanted pharmaceuticals. The program won a national award from the National Association of Clean Water Agencies (NACWA). The name of the program is "*No Drugs Down the Drain*" and this year, the program will be part of a one-year statewide event. Partners in the program include several cities and a neighboring utility. The *No Drugs Down the Drain* program takes advantage of existing programs that collect hazardous household waste (paints, batteries, etc.) to accept pharmaceuticals. It took several years to launch the program, in order to work through regulatory issues surrounding collection of waste pharmaceuticals. The program may help people feel that they are part of the solution. Public outreach efforts associated with the recently-launched Groundwater Replenishment System in Orange County has given a boost to positive perception.

4. Have you ever used Water Environment Federation (WEF) or Water Environment Research Foundation's (WERF's) resources on trace organic compounds or microconstituents, such as their website, Technical Practice Updates or Question & Answers? If so, was the information helpful? What other resources have been useful?

> Yes, they did come in handy when the AP stories broke. WEF sent a link to the site right when the stories were breaking, which was very helpful. The timeliness of WEF sending out the information made it much more useful. Additionally, Utility 4 has put together its own message points regarding microconstituents.

5. What, if any, are the principal concerns in your community related to trace organic compounds (e.g., human health concerns, environmental health concerns, other)? Please estimate the percentage of public concern associated with each.

> There have been a number of different concerns. There have been some environmental concerns regarding ocean aquatic organisms. Those interested in recycled water have focused on human health. There is probably more interest on the human health side of things versus the ecological side. The drugs in the water story created more health concerns rather than aquatic or environmental concerns.

6. Has your organization developed any partnerships to assist in communicating the risks associated with trace organic compounds (e.g., partnerships with local research organizations, medical doctors, consultants)?

Yes, the *No Drugs Down the Drain* program was first an initiative of Utility 4, a neighboring city, and a neighboring sanitation agency. Utility 4 has partnered with many research organizations in approximately 30 studies. They have not partnered with any other organizations formally on outreach, unless you count their support of this study for WERF. Utility 4 has several ongoing collaborations with Southern Nevada Water Authority, United States Geological Survey, several universities and others. Much of Utility 4's communications have been among scientists and information managers on this topic, rather than with the public. After their flatfish study is finished, they plan to sit down with a reporter to go over the results and explain the data.

Utility 4 has worked with the California Pharmacists Association and initially worked with the California Poison Control System to get the *No Drugs Down the Drain* program off the ground. (The California Poison Control System later withdrew its support for the program because they felt that the program does not support their overall mission. Since their main goal is to stop people from getting poisoned, some felt that putting something down the toilet might be more protective from a poison prevention perspective than disposing of it in the trash or storing these products).

Another example of Utility 4's leadership on pharmaceuticals in wastewater was the work they did on lindane, an insecticide that has been used in agriculture and also for the treatment of head lice in humans. Lindane is a neurotoxin that is readily skin-absorbed. Utility 4 lead the initiative to get pharmaceutical uses of lindane banned in California. Based on California's success, Mexico implemented a similar ban and Michigan is now close to a ban as well.

7. What tools, processes and approaches has your organization used to interact with the public? Describe the situations for which each approach has been used. Do you think that these methods attenuate public concern about trace organic compounds?

Utility 4 has done all of the standard things, including brochures, tours and talks. One thing that is unique perhaps is that Utility 4 purchases full and two-page ads in local newspapers during Earth week and other times. In fact, Utility 4 spends approximately $350,000 per year on communications and media. When the public sees something in the newspaper it has more credibility than when they see it in a brochure.

Utility 4 also has an active education program and even pays for buses to take children from their classrooms to the wastewater treatment plant. It was noted that this is anticipated to have a long-term and profound impact on how people think about this.

8. In your geographic area, have there been any changes in practices or programs to address trace organic compounds, such as monitoring programs, treatment studies or pharmaceutical take-back and disposal programs? Were any of these changes a direct result of public concern? Were any initiated by or supported by your organization?

All of the above have happened. Utility 4 has a pharmaceutical take-back program and has participated in numerous studies regarding to microconstituents. Utility 4's lab is developing internal capabilities to monitor EDCs and pharmaceuticals. NPDES discharge permits in Southern California are now requiring monitoring but Utility 4's permits do not yet include such a requirement, as none of them have been renewed in the last six months, when this requirement started to be required in the permits. Utility 4 is not opposed to monitoring and has developed analytical methods to sub-part per trillion (ppt) range (down to 0.5 ppt depending on the matrix). Utility 4 surveyed the literature, reviewed work on toxicological relevance of pharmaceuticals, assessed local regulations, and determine the availability of analytical standards and techniques to develop a list of compounds to monitor for. There was a wrinkle put in the last ocean permit requiring Utility 4 to take part in "special studies", and the effect of EDCs on flatfish was listed as an example. This study was proposed and is now a part of our "monitoring" requirement for our Ocean Discharge plant. Utility 4 has not seen any changes beyond the fact that there is now a statewide program for pharmaceutical take back.

9. Has your organization witnessed any changes in public involvement or perception regarding trace organics over time as a function of setting (e.g., during a drought, in response to an event, in response to the establishment of a program, etc.)? Please describe.

The media didn't take as active a role in 2008 when the AP report came out. It is thought that this was due to the thorough work of a prior reporter who had tracked the issue in 2006. When the AP article came out, the reporter's reaction was that the amounts were so small, the issue was not worth writing about as a human health concern in drinking water. Recycled water is a place where the rubber meets the road. Regarding drought, people's views are starting to change. Also, OCWD did a fantastic job on public outreach for its Groundwater Replenishment System, and changes in attitudes as a result of the outreach area being seen all around Southern California and even nationally. The California Regional Board is starting to pick up on this issue. Again the proactive stance taken by Utility 4 has made the organization a leader in this issue.

10. Please identify a local media report (article, radio or TV report) as an example of how the media presents information on this issue and how your organization is represented.

The 2006 article, also a DVD of local news reports will be sent to Elisabeth, to send to Linda.

11. Has your organization identified any communication deficiencies or needs (e.g., message content, style, trust, choice of media, etc.) that would improve the water industry's communication with the media and with the greater public on trace organic compounds/ microconstituents? Similarly, what approaches have been effective? Why?

> We also need a list of simple message points (8th grade education level) and a list of tips on how to deal with the media. For instance it would be helpful if someone would help to put the issue of detection into context. Is it useful to talk about 1 tablespoon in the Mississippi River rather than parts per trillion, which the public does not understand. Also, what do these amounts mean in terms of harm? Utility 4 thinks this would help reporters put things into simple, understandable terms. This could be a role for WEF or AWWA. WEF or another organization can help utilities deal with the media and also put the issue into terms that the public can understand. WERF has lots of information for industry and has published in journals but most people cannot understand the technical, scientific jargon.
>
> This would also help utilities with a credibility issue, because then they can reference an outside organization instead of presenting it as their own work and raising conflict of interest issues in the public's minds. When the public health agency and the water district aren't saying the same thing it erodes credibility. So there is a need to partner more with public health agencies or regulators, if possible.
>
> The public wants to know why we are learning about this now and they may feel something has been hidden from them rather than understanding improvements in analytical capabilities. EDCs aren't a new issue. In 1990, the issue of feminization of fish was being discussed in Europe. The media often presents this as a new issue, the scare-of-the-day. Perhaps there is a need for Journalism schools to train ethical reporters who don't sensationalize the issues to promote sales.

Interviews with Utility/Water District Teaming Partners
Utility 5

Date
July 31, 2008

Background
Rula provided a quick summary of the WERF study noting that the original scope was reduced. She explained that WERF wanted to first conduct a focused literature review on the knowledge and practice of public perception and risk communication generally, a literature review from other industries, and an analysis of media articles along with the input of utility partners. This is the current phase of work. We are interviewing ten different utilities. Rula noted that another critical component of the scope of work will be the utility's review of the draft report prior to sending it to WERF. Summary notes from this interview will be transmitted to back to Utility 5 for their review and comment. Final summary notes will be coded in an anonymous way when they are presented in the final report. Utility 5 is happy to participate in the project but that trace organics have not been much of an issue for them.

Questions and Responses
1. How important of a topic are trace organic compounds within your organization?

> Utility 5 has not done any real campaign with regard to trace organics locally. Pharmaceutical compounds is the term that is used because everyone feels they know what that is – microconstituents and trace organics are terms people don't understand. Utility 5 has also used the term "medicines" because it is more understandable to the broadest range of people and covers both prescription and nonprescription drugs.
>
> The issue has been important in terms of looking into the future to develop a more robust approach to water recycling issues. Two neighboring cities, on either side of Utility 5, have run into the issue of public acceptance of recycled water programs. One city was developing a recycled water program and in the other city, there was more fear of disease-causing organisms. So the issue is very important. This public consternation is creating more interest as a function of the movement toward recycled water and also because of the desire to protect receiving waters. A university is located near Utility 5 creates a very sophisticated as well as very eco-savvy constituency.

2. Does your organization have a designated spokesperson to answer questions from the public about trace organic compounds? Also, does your organization have a designated spokesperson to answer questions from the media about trace organic compounds?

> Utility 5 does not have a single designated person. Utility 5 usually ends up answering questions from the press but there is another person in the city manager's office who answers these questions from a water supply perspective. With the AP articles came out talking about pharmaceuticals in the water supply, people started asking about that.
>
> There hasn't been a heavy volume of calls on the topic, maybe one per week or so. Residents may call a few times per day, but the questions are principally about the medicine take-back program more than the "issue" of drugs in water. The medicine take-back program does have the impact of opening up a dialogue with the public. There are frequently short conversations generated by the interaction. Utility 5 had a medicine take-back program before the statewide *No Drugs down the Drain* program began, of which the city is a partner. Utility 5 is reluctant to use the word "drugs" because of illegal implications.

3. Has your organization prepared any outreach materials on the issue of trace organic compounds? If so, please describe these materials and how they were distributed.

> Yes, the program about medicine take-back was posted on Utility 5's website. Utility 5 has long been involved with regional issues as part of a county watershed program. They produced several white papers, one on the general subject of trace organics, one on triclosan and one on pharmaceuticals. Utility 5 can provide a link to these white papers, but they are probably posted on the watershed program's website.
>
> Utility 5 has a policy of not using triclosan. They prepared a fact sheet that explains to the public why they decided this and lists soap products that do not contain triclosan. This fact sheet is probably posted on Utility 5's website.

4. Have you ever used Water Environment Federation's (WEF's) or Water Environment Research Foundation's (WERF's) resources on trace organic compounds or microconstituents, such as their website, Technical Practice Updates or Question & Answers? If so, was the information helpful? What other resources have been useful?

> Yes, Utility 5 is reasonably sure that they have searched both websites. Another representative not on the teleconference has spent the most time on this topic and would have likely looked there.

5. What, if any, are the principal concerns in your community related to trace organic compounds (e.g., human health concerns, environmental health concerns, other)? Please estimate the percentage of public concern associated with each.

Most people are worried about drinking water but also recognize that a water problem in one location could become a problem in other water supplies. First, they are concerned about their own water, however. Utility 5 has explained to many people that their receiving waters are not a drinking water supply. Most people don't know where their water comes from but the result is that they are alert to every water problem. Utility 5 tries to explain basic water supply systems at every opportunity.

6. Has your organization developed any partnerships to assist in communicating the risks associated with trace organic compounds (e.g., partnerships with local research organizations, medical doctors, consultants)?

Utility 5 has coordinated with the local medical association chapter in the County, which has an environment subcommittee of doctors that are interested in environmental issues. They have designated an outreach program and plan to produce a couple of fact sheets. Utility 5 has gone to their meetings and talked about triclosan. Utility 5 also got a quote about triclosan from a member of this committee. So it may look like a partnership but it was a bit loose. Through the watershed program, Utility 5 has worked with other water districts and cities to develop white papers and participate in an emerging contaminant workgroup. A local association of water agencies is another vehicle for coordination on this issue between five adjacent Counties. The association became interested in the watershed program's white papers. A subcommittee of the association took some of the original white papers and adapted them. They may have prepared some original papers as well. They host local meetings and educate people about pollution prevention, among other things.

7. What tools, processes and approaches has your organization used to interact with the public? Describe the situations for which each approach has been used. Do you think that these methods attenuate public concern about trace organic compounds?

In addition to website and brochures, Utility 5 participates in events such as Earth Day and tables at different fairs. Their displays focus on the medicine take-back program or the alternative pesticide program. People are very interested in health and medicine and this helps to open up a dialogue with them. Utility 5 also provides people with information about alternative pesticides to use. People already get it that pesticides are toxic so there isn't a need to inform them of this issue. Utility 5's literature focuses on solutions that are less toxic or non-toxic.

8. In your geographic area, have there been any changes in practices or programs to address trace organic compounds, such as monitoring programs, treatment studies or pharmaceutical take-back and disposal programs? Were any of these changes a direct result of public concern? Were any initiated by or supported by your organization?

> Most of the programs have been initiated because Utility 5 was concerned that surface water was being contaminated by the community, not because residents were up in arms about trace organics. It is a combination of staff recognition of a potential problem and state requirements. The State has required monitoring for pesticides and may have requirements to act (e.g., to develop product substitutions) in order to reduce discharges to the sanitary sewer and storm system. The state may require some development of monitoring, outreach and control but not with great specificity. So the City submits their annual report with what they did to reduce discharges to the sanitary sewer system and storm water system. So depending on how microconstituents are defined, there may be requirements to monitor or to act.

9. Has your organization witnessed any changes in public involvement or perception regarding trace organics over time as a function of setting (e.g., during a drought, in response to an event, in response to the establishment of a program, etc.)? Please describe.

> It is hard to identify why things have changed – probably in response to all of these things – but there is a greater public awareness of water quality, especially during the last decade. Public awareness has increased including staff awareness, as part of the public. Utility 5 does not think this increase can be attributed to any one thing. Definitely the press articles and medicine take-back program have had an impact. These are probably the two biggest reasons.

10. Please identify a local media report (article, radio or TV report) as an example of how the media presents information on this issue and how your organization is represented.

> It is hard to remember. The recent AP series of articles highlighted concerns on PPCPs. Utility 5 wasn't represented in the AP article but talked with the local newspaper afterwards. The local media are pretty good about getting things and did not appear motivated to try to scare the public. The issue got fairly good press.

11. Has your organization identified any communication deficiencies or needs (e.g., message content, style, trust, choice of media, etc.) that would improve the water industry's communication with the media and with the greater public on trace organic compounds/ microconstituents? Similarly, what approaches have been effective? Why?

The public doesn't understand why there isn't better data. It sounds lame to the public that we haven't tested for various compounds just because it is expensive or that we don't know how to detect different compounds. When you step away from this you quickly see how it sounds silly that we let people take these drugs without having a good testing program about how much gets into water. People can understand that we aren't doing a more holistic job of testing – for example, that we have never tested in receiving waters for these compounds or worse, have never tested their drinking water.

Interviews with Utility/Water District Teaming Partners
Utility 6

Date
July 31, 2008

Background
Rula provided an overview of the project background and scope of work, which was reduced since the initial proposal and letters of in-kind commitment were submitted to WERF. The four main tasks include a focused literature review, a review of literature published on risk communication by other industries, an analysis of media coverage and preparation of a draft and final report. As part of the media analysis task, we are also conducting interviews with nine or ten different utilities. At the end of the call, we will send the notes to Utility 6 for review. Utility 6 will send us back any comments or corrections to the transcript. Utility 6 will also have an opportunity to review the draft report, which we will be able to send out in mid- to late-August.

Questions and Responses

1. How important of a topic are trace organic compounds within your organization?

Very important. This issue is at the heart of credibility of water quality. Regarding the aspects of trace organics that Utility 6 has focused on (e.g., source control? water reuse?), there is not yet any pharmaceutical take-back program. Highly treated wastewater is reused for non-potable purposes, either for reinjection as a seawater barrier or for industrial reuse, so there are no direct potable reuse concerns.

2. Does your organization have a designated spokesperson to answer questions from the public about trace organic compounds? Also, does your organization have a designated spokesperson to answer questions from the media about trace organic compounds?

Utility 6's representative is the designated spokesperson. Unfortunately, he has only been at Utility 6 for two months and has not fielded a lot of calls from the public. It does not seem like it's too much different of a climate than the utility where Utility 6 representative worked previously. At the previous utility, the key issues involved 1,4-dioxane and NDMA detections in 2000 to 2001. More recently, the AP article caused a bit of media interest at the previous utility. Their advanced treatment system was able to remove pharmaceuticals and other compounds mentioned in the AP article. However, this utility still did not have a point of reference for health effects and needed this from toxicologists.

3. Has your organization prepared any outreach materials on the issue of trace organic compounds? If so, please describe these materials and how they were distributed.

Utility 6 has not prepared any of these materials yet. On the list of materials that they plan to prepare is a fact sheet on trace organics and other information on the website. The information could be posted online or distributed in take-home packs. There is an outreach program at Utility 6 for the community, known as the Speaker's Bureau Program. One of the topics offered is water quality. To set up a talk, Utility 6 would send a letter out to the Speaker's Bureau organizations saying that the following speakers were available and enclose a pamphlet advertising the event.

4. Have you ever used Water Environment Federation (WEF) or Water Environment Research Foundation's (WERF's) resources on trace organic compounds or microconstituents, such as their website, Technical Practice Updates or Question & Answers? If so, was the information helpful? What other resources have been useful?

The representative has not used these resources at Utility 6 yet but will soon.

5. What, if any, are the principal concerns in your community related to trace organic compounds (e.g., human health concerns, environmental health concerns, other)? Please estimate the percentage of public concern associated with each.

Most of the concern (>95%) is related to human health issues related to drinking water. Utility 6 does not have examples. In addition, there are also environmental groups who are concerned about fish, etc.

6. Has your organization developed any partnerships to assist in communicating the risks associated with trace organic compounds (e.g., partnerships with local research organizations, medical doctors, consultants)?

Utility 6 will probably partner with another utility since Utility 6 is aware of their capabilities. Utility 6 may also partner with research organizations if the opportunity arises. Utility 6 is strongly pro-research and has supported WERF, AwwaRF and other applied research projects in the past. Utility 6 has also worked with a number of different universities.

7. What tools, processes and approaches has your organization used to interact with the public? Describe the situations for which each approach has been used. Do you think that these methods attenuate public concern about trace organic compounds?

These have already been mentioned and include a fact sheet, website and speaker's bureau topic on trace organic compounds.

8. In your geographic area, have there been any changes in practices or programs to address trace organic compounds, such as monitoring programs, treatment studies or pharmaceutical take-back and disposal programs? Were any of these changes a direct result of public concern? Were any initiated by or supported by your organization?

> Utility 6's representative is taking over the program from an engineer who was acting as the point of contact on public and media issues. Utility 6 decided that they needed to do this and support the overall program.
>
> There is a regional pharmaceutical take-back program that other adjacent utilities and cities are involved with. Utility 6 has conducted some monitoring for trace organic compounds but Utility 6's representative is not aware of the details.

9. Has your organization witnessed any changes in public involvement or perception regarding trace organics over time as a function of setting (e.g., during a drought, in response to an event, in response to the establishment of a program, etc.)? Please describe.

> Just the AP story reaction. The representative from Utility 6 was not at the utility at the time but imagines the response would have been similar to that at his previous job at another utility.

10. Please identify a local media report (article, radio or TV report) as an example of how the media presents information on this issue and how your organization is represented.

> Utility 6 representative is too new to the organization to answer this question.

11. Has your organization identified any communication deficiencies or needs (e.g., message content, style, trust, choice of media, etc.) that would improve the water industry's communication with the media and with the greater public on trace organic compounds/ microconstituents? Similarly, what approaches have been effective? Why?

> One example is the need to establish safe thresholds for various constituents so that utilities can report to the public that the detected levels are well below or in the same magnitude of the concern. There is a study ongoing right now with Shane Snyder, Orange County Water District and a group from Australia to look into this. It would be nice to find a way to compare levels of these constituents in water to a safe dose. The medical industry could weigh in if the levels of these compounds are not a health concern. There may need to be more human health research/toxicology studies before we can say that.
>
> Regarding terminology, Utility 6 has referred to these compounds as trace levels of pharmaceuticals and personal care products. Other utilities have been calling them medicines.

Interviews with Utility/Water District Teaming Partners
Utility 7

Date
August 1, 2008

Background
Rula provided an overview of the project. Malcolm Pirnie described the WERF Workshop on Trace Organic Compounds: Mapping a Collaborative Research Roadmap. One common theme that emerged from this workshop was the importance of risk communication and the need to re-focus the RFP on risk communication. The scope of work now includes harvesting the literature and seeing if there is something to learn from other industries as a Phase I before embarking into focus groups and additional surveys. After the focused literature search, we will conduct a media analysis and review. Part of this task is conducting interviews with nine utility partners. The project Issue Area Team (IAT) specifically recommended Utility 7 as an interview participant as they are a smaller utility. Rula noted that their involvement will be particularly helpful.

Questions and Responses

1. How important of a topic are trace organic compounds within your organization?

The topic is important from the standpoint of the public as it can feed into public fear. It is an important topic for public drinking water across the U.S. including surface waters and groundwater. The general public is concerned about pharmaceuticals in source waters. The public thinks about pharmaceuticals and does not use the term "trace organics". At Utility 7, the phrase "trace organics" relates to volatile organic compounds (VOCs) or disinfection byproducts (DBPs). Neither VOCs nor DBPs have been a public issue. Other issues, such as lead, are a higher priority for Utility 7. The topic of communication is becoming more important as we learn more and detection methods improve. For example, Utility 7 tried to relate the amount of caffeine found in water to the amount in a cup of coffee. It is difficult for callers to understand why Utility 7 is not testing for these constituents (i.e., pharmaceuticals that were the topic of the AP report, particularly when other nearby utilities are. Utility 7 has explained that they wouldn't know what to do with the test results. They wouldn't know if detection equated to a health hazard or not. They couldn't recommend other drinking water sources to the public.

2. Does your organization have a designated spokesperson to answer questions from the public about trace organic compounds? Also, does your organization have a designated spokesperson to answer questions from the media about trace organic compounds?

It depends on the question. Utility 7 serves 110,000 people and does not have large utility staff or a designed public relations specialist. Questions would be answered by their General Manager or by the Chief of Operations in Water Quality or by the Laboratory. Media questions would go through the General Manager. Calls to the general office would have been directed to any of these three representatives.

3. Has your organization prepared any outreach materials on the issue of trace organic compounds? If so, please describe these materials and how they were distributed.

No, Utility 7 has not prepared any outreach materials.

4. Have you ever used Water Environment Federation (WEF) or Water Environment Research Foundation's (WERF's) resources on trace organic compounds or microconstituents, such as their website, Technical Practice Updates or Question & Answers? If so, was the information helpful? What other resources have been useful?

No, but Utility 7 has used AWWA and AWWARF as well as the State Department of Natural Resources. AWWA provides email alerts and contacts for further information. If WEF has similar information on their website, it would help to distribute it to people by email.

5. What, if any, are the principal concerns in your community related to trace organic compounds (e.g., human health concerns, environmental health concerns, other)? Please estimate the percentage of public concern associated with each.

People have almost all expressed human health-related concerns. People don't want to be involuntarily exposed to risks and want to know what is in the water. Environmental health issues have not really been raised by the public.

6. Has your organization developed any partnerships to assist in communicating the risks associated with trace organic compounds (e.g., partnerships with local research organizations, medical doctors, consultants)?

Not in regards to this issue, no. There haven't been any partnerships developed at this point.

7. What tools, processes and approaches has your organization used to interact with the public? Describe the situations for which each approach has been used. Do you think that these methods attenuate public concern about trace organic compounds?

Utility 7 has not communicated with the public on trace organics/ pharmaceuticals, aside from answering callers' questions (see response to questions 1 and 8). Utility 7 has included bill inserts about lead and copper and provided an annual Water Quality Report. At this point, there haven't been a great number of calls on the topic. The number of calls is not logged in a centralized system, though some departments might write down names and phone numbers. Utility 7 has talked about tracking calls in the past and does track calls about other information, such as water main breaks. There is some interest on the part of Utility 7 to have such a centralized system.

8. In your geographic area, have there been any changes in practices or programs to address trace organic compounds, such as monitoring programs, treatment studies or pharmaceutical take-back and disposal programs? Were any of these changes a direct result of public concern? Were any initiated by or supported by your organization?

Utility 7 has had a household hazardous waste collection facility since 2006. Funds were made available to the City Health Department to add a pharmaceutical take-back program with one collection day in 2007 and one in Spring 2008. This collection has been very popular and well-received, especially among the elderly. People want to do the right thing.

There is no mandate for monitoring. EPA collected some samples from Utility 7's wastewater and sampled for about 200 different chemicals, as part of a separate study. Perch populations are declining in nearby surface waters because 98% of them are now female. People believe that this is from exposure to endocrine disrupters. There is a growing concern and information in the media about trace organics showing up in wastewater. It is not known what changes will be needed. A city directly north of Utility 7 has voluntarily conducted sampling of source waters, finished waters and distribution system. Out of 70 to 100 different analytes, they found five compounds in the source water, two in the finished water and nothing detectable in the distribution system. They are voluntarily testing but Utility 7 is not, as there is no science to interpret the results or assess the impact on people and their families. Sampling may place fear in people and/or steer them towards bottled water, which they think is safer or cleaner, though it probably has not been tested for these compounds either. To cover the costs of additional sampling, Utility 7 might have to raise their rates and hire another laboratory technician. The other utility nearby is larger and has more funding capabilities compared with Utility 7. By monitoring, the neighboring utility can say they are doing something about the situation. Smaller jurisdictions need to have more assistance.

(Response to question #8, continued...)

Not sampling is problematic too, because people want to know why Utility 7 is not testing for these compounds and think that Utility 7 may be hiding something. It is hard for the public to understand why their drinking water utility is not testing for something that is in the news. The Superintendent of Wastewater made a presentation about water infrastructure recently at a community event. During the Q&A, he was asked if Utility 7 is testing for pharmaceuticals in water. Utility 7 has told the public that Utility 7 is not required to test and doesn't know the effects of these compounds but this doesn't make people feel better either. Utilities are "damned if they do [sample] and damned if they don't".

This issue is out in the news media but no one has answers. Utility 7 has developed some communication tools to communicate the minute amounts detected – for example, the amount of water you'd have to drink before getting enough caffeine to equal one cup of coffee. After receiving the AP article news alert from AWWA, Utility 7 called to ask if they had any information but didn't get anything back from AWWA. There needs to be a pamphlet or public information document to make available to staff on Q&A. Maybe this can come out of the WERF project. This issue is not going away.

9. Has your organization witnessed any changes in public involvement or perception regarding trace organics over time as a function of setting (e.g., during a drought, in response to an event, in response to the establishment of a program, etc.)? Please describe.

Utility 7 received a few calls after the AP article, which diminished to one or two calls per month. This does not include the number of calls received by the Laboratory (see response to question #2).

10. Please identify a local media report (article, radio or TV report) as an example of how the media presents information on this issue and how your organization is represented.

The media published the AP article but this did not mention Utility 7 specifically. Utility 7 has not responded by sampling since the health effects are still unknown (as described in the response to question 9). After health effects are known, it may take 20 years for EPA to promulgate rules and establish levels of effects. The problem will not go away anytime soon.

11. Has your organization identified any communication deficiencies or needs (e.g., message content, style, trust, choice of media, etc.) that would improve the water industry's communication with the media and with the greater public on trace organic compounds/ microconstituents? Similarly, what approaches have been effective? Why?

The problem we run into is that everything is sensationalized in the media, which creates public fears. People are concerned and start to doubt public agencies and have safety concerns. Getting the good word out about drinking water is difficult. We need to put out an alternative point of view, perhaps through WEF or AWWA as a contact for local news media to restore public confidence, which can be slow. It has taken years to recover public confidence from the cryptosporidium outbreak in Milwaukee.

Another challenge is the lack of scientific knowledge in the general community. It is hard to convince people that water is safe. They may say, 'Why should I trust you? – you are not even testing for these constituents. I trust AP more than you as a local politician'. They may feel they are just getting a song-and-dance around the issue. Water and wastewater treatment agencies may not be the most credible sources on water quality and we need others' help. WEF and AWWA could help utilities by preparing positive information and messages that could be transmitted to utilities before a media campaign is anticipated to help them with something to say to callers. Topics could include: What are the messages? Who are the experts?

The public needs to know how to put the issue in context. There is more trust in bottled water as a purer source of drinking water than tap water. Utility 7 has explained to callers that bottled water quality is not regulated and that water is not pure; it is a universal solvent. Some water basics could help the public understand these concepts.

Interviews with Utility/Water District Teaming Partners
Utility 8

Date
August 4, 2008

Background
Elisabeth provided a brief overview of the project, emphasizing that the focus is to examine the public perception of trace organic compounds. Elisabeth discussed the narrowed scope of work from the project that was originally supported by Utility 8. WERF wanted first to obtain a focused literature review on the knowledge and practice of public perception and risk communication, a literature review from other industries, and then obtain an analysis of media articles along with input from utility partners. Utility partners will provide input through theses interviews and will also review and comment on the draft report. A summary of the notes from this interview will be transmitted to back to Utility 8 for review and comment. Utility involvement will help to shape the final report.

Questions and Responses

1. How important of a topic are trace organic compounds within your organization?

> Utility 8 likes the term trace organics. Utility 8 has not embraced the term microconstituents like other water districts have in the past. They are sensitive to the idea of using neutral language versus a term like "contaminants" but try to avoid euphemisms. Utility 8 has been adaptable with other terms and hasn't locked into any specific term. They work with the media and use their language. The World Health Organization (WHO) appears to be using the term compounds or contaminants of emerging concern. Utility 8 noted that in a recent meeting of the Singapore Independent Science Panel, Dr. Ong Choon Nam and Dr. Joan Rose, both international experts, used the term compounds of emerging concern. This may be growing in acceptance but may not carry over to dialogue with the public. Utility 8 noted that within the industry there is more of a tendency to be frank and less precise with terms. Almost all of the compounds (except perchlorate) are trace organic compounds. Dr. David Sedlak and Dr. George Tchobanoglous have used the term "pharmaceutically active compounds". The term "trace organics" is an older term. Utility 8 has rolled with whatever the current term seems to be and they are comfortable with a lot of terms – haven't tried to push the industry one way or another.
>
> The topic is very important. The Board is very aware of the issue, dating back to a study on a nearby river system. The Board realizes they are in the spotlight with their advanced treatment and reuse operations and is supportive of Utility 8's thinking about these issues and having answers to the questions that come their way. About half of the people who go through the Utility 8's tour bring up trace organic

Response to Question 1, cont'd...

compounds. On a staff level Utility 8 has been involved in the topic, looking at treatment effectiveness, the nearby river system, pilot- to full-scale research for advanced oxidation and reverse osmosis, testing protocols, treatment processes and outreach. Utility 8 has tried to look at how well will the process work and how well natural barriers in the groundwater system assist with the process. Utility 8 is actively involved in research on the topic.

2. Does your organization have a designated spokesperson to answer questions from the public about trace organic compounds? Also, does your organization have a designated spokesperson to answer questions from the media about trace organic compounds?

The answer to this varies on a case-by-case basis, depending on how technical the question appears. All questions from the public and media go through the Public Affairs Department. The public affairs coordinator has a team of two or three people who can address the questions. The type of question and a persons' availability plays into it as well. To some extent every single tour addresses these questions so whoever conducts the tours may field a question(s) about the issue. In general, such questions go to the General Manager or one of three other staff members. Public affairs department staff will return calls from the public and seek answers to questions if they determine they do not have enough information to answer them.

Sometimes a Board Member or former Board Member might be a spokesperson for the advanced treatment system. For example, a former Board Member is now a State Assemblyman but still speaks about the advanced treatment system.

3. Has your organization prepared any outreach materials on the issue of trace organic compounds? If so, please describe these materials and how they were distributed.

These topics are addressed as they relate to the advanced treatment system's effectiveness and multiple barriers for removing these compounds (both microbial and chemical) from water. CECs are not addressed in a stand-alone document.

4. Have you ever used Water Environment Federation (WEF) or Water Environment Research Foundation's (WERF's) resources on trace organic compounds or microconstituents, such as their website, Technical Practice Updates or Question & Answers? If so, was the information helpful? What other resources have been useful?

No, Utility 8 has not used WEF and WERF materials. They have used AwwaRF materials – AwwaRF tried to be available when the AP articles came out, and made findings from recent research available. This was helpful. A number of Utility 8 staff have been active in WEF but not as active with WERF. Utility 8 received WERF funding for long-term testing/fish bioassays but are less plugged in to the tools available from WERF.

5. What, if any, are the principal concerns in your community related to trace organic compounds (e.g., human health concerns, environmental health concerns, other)? Please estimate the percentage of public concern associated with each.

Both human health and environmental health issues are a concern, because of the coastal community. But there is more concern about human health because of the advanced treatment system. Utility 8 has had a very active coastal water quality contingent with community groups and an awareness of these compounds in the marine environment. There is less awareness and less concern over the nearby river system, which is mostly wastewater and is less treated than their reclaimed water. Utility 8 noted that there could be more concern about the river water quality in the future which would relate mainly to the issue of human health concerns, though there are some threatened or endangered species in the river, as the river is a drinking water supply for several million people.

6. Has your organization developed any partnerships to assist in communicating the risks associated with trace organic compounds (e.g., partnerships with local research organizations, medical doctors, consultants)?

Yes. Utility 8 has stressed partnering in a number of ways and venues. Utility 8 has had a number of Scientific Advisory Panels which had some physician and community group participation. A local physician and environmentalist, and the Executive Director of a coastal community organization are examples. They track the treatment system developments and give feedback to Utility 8 from the public. Because of the nature of the water reclamation system, Utility 8's recently former Public Relations Director was not just talking to people but listening to people. He continually adjusted the message based on the feedback he was getting. The public was a partner. And to be frank, the regulators are partners as well. All the meetings with the regulators created an environment that when emerging contaminants were found, a representative from the State Department of Health sat with them during media presentations. The State Department of Health was in agreement with what they said so there were partners in presenting consistent information to the public.

Also, Utility 8 has worked with the a local watershed group comprised of five water agencies, three counties and others to provide a partnership focus that attempts to balance diverse interests. That group's Executive Director, formerly with the State Water Resources Control Board, has encouraged a watershed-wide look at water quality.

7. What tools, processes and approaches has your organization used to interact with the public? Describe the situations for which each approach has been used. Do you think that these methods attenuate public concern about trace organic compounds?

Utility 8 has been using everything from focus groups to surveys on their treatment systems to ad-hoc responses to NDMA issues. When Utility 8 encountered NDMA, they went out to talk to the media on the advice of their communications director. Utility 8 broke the story about NDMA (something that wasn't customarily done), approached the media, told their story and actually got a lot of flack from other agencies. However, breaking the story themselves did a lot to bolster public confidence in Utility 8 and built trust in Utility 8 to address the problem when they encountered it. Utility 8 has prepared literature about their advanced treatment system. For example, literature talks about reverse osmosis (RO) as an effective barrier against salts and pharmaceuticals.

8. In your geographic area, have there been any changes in practices or programs to address trace organic compounds, such as monitoring programs, treatment studies or pharmaceutical take-back and disposal programs? Were any of these changes a direct result of public concern? Were any initiated by or supported by your organization?

Utility 8 hasn't hosted any pharmaceutical take-back programs but they have been actively involved in getting information out to the public, with a neighboring utility about proper disposal of pharmaceuticals, solid waste and hazardous waste. Utility 8 is currently involved in the statewide No Drugs down the Drain program but in a support role. Utility 8 is conducting monitoring of their 70 MGD system. They are obligated by permit to test for various CECs, including a number of pharmaceuticals. They monitor for several others based on the advice of an independent advisory panel and conduct other monitoring to determine treatment effectiveness. Utility 8 has worked closely with a number of researches such as Jörg Drewes, David Sedlak and Shane Snyder regarding their treatment systems.

9. Has your organization witnessed any changes in public involvement or perception regarding trace organics over time as a function of setting (e.g., during a drought, in response to an event, in response to the establishment of a program, etc.)? Please describe.

Historically Utility 8's attention has been reactive because the public reads something and calls them or the media calls. These days, Utility 8 is becoming more of a source of information. The new plant was brought on-line in January and they have had questions are coming from surrounding counties and cities. The public hasn't been as concerned. They are concerned about Utility 8's ocean discharge and they are concerned about the ocean impacts of brine outfalls from desalination plants. But this hasn't grown to be a big group throughout the county. Hopefully there is less attention because Utility 8 has been being more proactive in getting out information.

10. Please identify a local media report (article, radio or TV report) as an example of how the media presents information on this issue and how your organization is represented.

The media typically calls Utility 8. For the most part, they are in the same boat as other utilities. There was an article in the NY Times about groundwater recharge following the AP story that generated some questions. Tours are very helpful because questions get answered on the spot. People leave with fewer questions. Utility 8 has a constructive relationship with the local news. One journalist has been supportive of the "just get over it" message to the public about water reuse and sarcastic about the opponents of reuse. Others try to be scientific writers. The area has one radio show that has bashed water reuse. Utility 8 has offered for them to come out to the plant but the radio show hosts are just looking for a laugh and like to play up the "yuck" factor. NPR called a lot of attention to the issue. Another local radio station has a reporter who is very open to what Utility 8 is doing.

Most reporters have understood the materials that Utility 8 has provided. A local paper had a front page article called Prescription for Worry. During the time that the AP came out with their story, the Eliot Spitzer story broke and pushed the issue off the front page – apparently the media loves sex more than chemistry. The issue tends to be presented in a more sensationalized manner in Latino or Spanish television, perhaps in response to or creating a greater mistrust of water supplies.

11. Has your organization identified any communication deficiencies or needs (e.g., message content, style, trust, choice of media, etc.) that would improve the water industry's communication with the media and with the greater public on trace organic compounds/ microconstituents? Similarly, what approaches have been effective?

Groups like WEF, WERF, AwwaRF and WateReuse Foundation could do a better job of being perceived as an objective source of information. This is the reason behind AWWA changing its name. We have a long ways to go before we present an integrated water industry perspective. The most effective approach is to be honest. The first reaction from a public agency and perhaps from the water associations is to cringe when the questions start. Utilities can approach it proactively and try to play devil's advocate to get prepared for questions. This puts everyone at ease.

It also helps to have and give benchmarks, such as Drinking water equivalency levels (DWELs) from a recent AwwaRF study so that people have a sense of the health effects and concern may be mitigated. Utility 8's former communications director was good about showing a voluntary risk comparison to give perspective. When NDMA was an issue, he used the example of how much NDMA is in foods that are voluntarily consumed. (DWELs are a way to put drinking water concentrations in perspectives. It is a similar approach to the one OEHHA uses, to come up with a de minimus risk level and include appropriate margins of safety.

Interviews with Utility/Water District Teaming Partners
Utility 9

Date
August 7, 2008

Background
Elisabeth provided a brief overview of the project emphasizing that the focus is to examine the public perception of trace organic compounds. Elisabeth discussed the narrowed scope from the project that was originally supported by Utility 9. She explained that WERF wanted to first conduct a focused literature review on the knowledge and practice of public perception and risk communication, a literature review from other industries, and conduct an analysis of media articles along with the input of utility partners. Utility partners will be involved in the interviews and also be asked to review and comment on the draft report. The Malcolm Pirnie team will send summary notes from the interview back to Utility 9 for review and comment. When written up for the final report, all information will be presented anonymously, as Utility A, Utility B, etc. Ed Means and Linda Macpherson then began asking the questions.

Questions and Responses

1. How important of a topic are trace organic compounds within your organization?

> We first explored what Utility 9 uses as a term(s) to refer to the topic. Utility 9 was part of the discussion with WERF on which term to use. Utility 9 generally uses "Emerging contaminants" as their preferred term because it is one of the longest used terms. "Compounds of emerging concern" or "Compounds of potential concern" is another term that was often used. Utility 9 actually suggested the phase "Compounds of potential concern" which was initially embraced by the WEF Community of Practice though the preferred term was changed later. EPA has used "compounds of emerging concern" for a while. Most people understand these terms more than those currently in use which seem more like semantics or mincing words. It is important that the public has a sense of these items as something of concern but it is true that we don't want to heighten concerns if we don't know the impact of these contaminants.
>
> Utility 9 believes the issue is important with the principal support coming from the wastewater side. There is some concern on the water side to not talk too much about it due to concerns that the public make an incorrect judgment that there is something wrong with their drinking water supply (which is snow melt). Utility 9's drinking water supply is pristine and is not downstream of wastewater treatment plants. Because Utility 9 is a water and wastewater utility, it is important to be really careful (cont'd...)

(Response to Question 1, cont'd...)

about how messages are presented to the public. Water reuse has become a bigger need at a regional level and there is concern that the concerns about trace organics might impact the acceptance of reuse programs. For example, Utility 9 is working on the statewide campaign *No Drugs Down the Drain*. Among several graphics, the campaign uses an illustration of a fish wearing a gas mask. A local water association expressed concern that the image was alarmist and could impact recycled water programs. Other agencies that have been doing water recycling at a larger scale have used this graphic for more than two years and have seen no problem with it. It shows that there are different sensitivities and different media markets across California.

2. Does your organization have a designated spokesperson to answer questions from the public about trace organic compounds? Also, does your organization have a designated spokesperson to answer questions from the media about trace organic compounds?

There is a designated spokesperson from Utility 9's media office that handles all calls from the media. Sometimes calls will be kicked back to one of three other people. It depends on the question. For public inquiries that are created through a hotline of public service announcements, they may be answered by anyone in the section. Utility 9 gets a lot of questions about pharmaceuticals. Who answers the call really depends on how detailed the question and the nature of the question.

3. Has your organization prepared any outreach materials on the issue of trace organic compounds? If so, please describe these materials and how they were distributed.

The *No Drugs Down the Drain* campaign with the "gas mask" image was already discussed. Utility 9 has also highlighted research on triclosan. Utility 9 has phased out use of triclosan, an approach that has been embraced by others (including two neighboring cities). The issue is addressed on Utility 9's website by a white paper. Sometimes Utility 9 gets calls about triclosan. The information is distributed to the public in bill inserts. Proper drug disposal days are noted on calendar listings, newspaper ads, public service announcements and flyers, which can generate phone calls from the public. Utility 9 conducted a wastewater source study in conjunction with a local non-profit to look for triclosan, phthalates, bisphenol-A and nonylphenol. Results were publicized in a report posted on their website and by the Environmental Working Group (EWG). Utility 9 and EWG conducted a press conference which gained front-page news, television coverage, etc. Results will soon be published in the journal Science of the Total Environment.

4. Have you ever used Water Environment Federation (WEF) or Water Environment Research Foundation's (WERF's) resources on trace organic compounds or microconstituents, such as

their website, Technical Practice Updates or Question & Answers? If so, was the information helpful? What other resources have been useful?

> Utility 9 used to be a WEF member and have access. There are only a few people with paid membership at Utility 9. One person was trying to access the TPUs and was not able to. WEF did send out talking points by email in preparation for the AP articles. Utility 9's most useful resource is a network of individuals and organizations who are interested in this issue as well as the articles found online. Conferences are also good places to get information and copies of presentations. One representative at Utility 9 did her thesis on this topic so she has materials and contacts from this work as well.

5. What, if any, are the principal concerns in your community related to trace organic compounds (e.g., human health concerns, environmental health concerns, other)? Please estimate the percentage of public concern associated with each.

> There is a lot of confusion among residents about the issue. The region has an environmentally-oriented press. In general, people want to be told what they should and shouldn't buy. They want to know what they can do in their homes to reduce concentrations of trace organics. There is some data out there but no comprehensive source of information. There is also a lot of confusion about whether or not the presence of drugs in water will harm people. Since these compounds are known to harm critters, most people make that leap to human health impacts. "If it hurts frogs, it can harm people." When the AP article came out there were probably more calls going to the water side of Utility 9. Those calls would likely have gone to Utility 9's drinking water department, who developed the key responses. Wastewater did get some calls also.

6. Has your organization developed any partnerships to assist in communicating the risks associated with trace organic compounds (e.g., partnerships with local research organizations, medical doctors, consultants)?

> The journal article was mentioned previously. Utility 9 has partnered with two non-profits. One non-profit helped Utility 9 get the word out about and conduct pharmaceutical take-back events. Regarding triclosan, Utility 9 has worked behind the scenes with several entities, including a local water utility association, who is going to draft a letter to all member agencies regarding *No Drugs down the Drain*. Utility 9 is also participating in a regional monitoring program. They are conducting monitoring for priority pollutants and conducting some studies for non-priority pollutants. In the coming year, they plan to focus on pharmaceuticals. Triclosan is also of interest, as well as poly-brominated diphenyl ethers (PBDEs) and others.

7. What tools, processes and approaches has your organization used to interact with the public? Describe the situations for which each approach has been used. Do you think that these methods attenuate public concern about trace organic compounds?

> Utility 9 uses several approaches to interact with the public regarding emerging contaminants. First, Utility 9 publicizes information in their bill insert newsletter. They bill every other month and include in each newsletter an article related to wastewater pollution prevention ("The Pollution Prevention Corner"). P2 Corner is well-received by the public and typically generates phone calls to the wastewater department. Utility 9 publicizes their medicine and mercury thermometer exchange events. Another approach is providing information on Utility 9's website, as mentioned below. Finally, they try to have direct interaction with the public through presence at Earth Day events, other public fairs and through treatment plant tours. At public events, they typically provide outreach materials about home toxics, medicine disposal, mercury, less-toxic cleaning and gardening products. They offer wastewater treatment plant tours and discuss issues like emerging contaminants. In the context of seeing a wastewater treatment plant, the public really "gets" that it's not easy for them to change technologies to address a given emerging contaminant. Utility 9 also gives away gardening and household cleaner guides, sponges with information on non-toxic cleaning and other information at the tour. The public they interact with are generally happy to learn how to reduce their own exposure to potential toxins/emerging contaminants and also to protect receiving waters.
>
> The response to the AP article was pretty typical and illustrates the traditional way of responding to an issue raised by others. Perhaps responsiveness would have been improved by more proactive development of talking points ahead of time. There wasn't a coordinated approach regarding talking points between Utility 9's water and wastewater. The drinking water side could have talked more about all the work that Utility 9 is doing on the wastewater side -- collecting unused drugs and preventing pollution. Utility 9 did a study in 2002 and did not find any of the compounds that they looked for. This was a means to allay public fears. The AP articles were focused mainly on drinking water, so there was not a lot of focus on where the wastewater was discharged (to receiving waters).

8. In your geographic area, have there been any changes in practices or programs to address trace organic compounds, such as monitoring programs, treatment studies or pharmaceutical take-back and disposal programs? Were any of these changes a direct result of public concern? Were any initiated by or supported by your organization?

Utility 9 started a pharmaceutical take-back program in 2005 so, in this sense, they were ahead of the curve. Utility 9's activities didn't really change because of AP articles or general knowledge of compounds of emerging concern. In fact, their programs have been helped by the coverage. Implementing a pharmaceutical take-back program is the low-hanging fruit for wastewater. On the water side, Utility 9 has not made any decisions to increase monitoring. Also, the regional monitoring program has conducted several studies on emerging contaminants. The program is funded by an umbrella group for wastewater agencies and the Regional Water Quality Control Board.

9. Has your organization witnessed any changes in public involvement or perception regarding trace organics over time as a function of setting (e.g., during a drought, in response to an event, in response to the establishment of a program, etc.)? Please describe.

There is some ebb and flow to the public attention as a direct function of media coverage. Utility 9 has seen more consistent concern recently. Interest and awareness of pharmaceuticals in water has grown over the past three years. As Utility 9 attends events, they have more people coming and participating in them. Some say there is a need to say something three times before seeing any change or action. Maybe Utility 9 is seeing that now in the community. One representative was brought on board specifically to do more outreach and community action. Utility 9 goes to Earth Day and other events and publish online and print guides about how to make one's own household non-toxic cleaners. Over time, it seems like the word is getting out. The number of hits on their website is increasing, indicating a growing public response.

The online guides and outreach articles have been helpful in spreading the word. There is a cleanup guide for use at home, as well as a grow-it guide, which focuses on non-toxic products in gardens. This is a stormwater issue but garden products might also go down the drain. Utility 9 references these documents in their newsletter. The next newsletter will feature nanoparticles, and will explain why Utility 9 is not giving rebates on Samsung nanoparticle washing machines, as they don't want water going down the drain that contains silver nanoparticles. Utility 9 is ahead of the curve calling attention to the use of nanoparticle-containing consumer products.

10. Please identify a local media report (article, radio or TV report) as an example of how the media presents information on this issue and how your organization is represented.

The AP articles were a great example of media coverage where there wasn't a lot of local information incorporated into the article. It left a lot of agencies scrambling to respond. There was some incorrect information. For example, AP reported that local drinking water contained estrogen. This was untrue and retracted the next day but the placement of the information left many without knowledge that this fact had changed from the original story. It would have been better if reporters had worked with local officials or utilities before publishing the AP story.

11. Has your organization identified any communication deficiencies or needs (e.g., message content, style, trust, choice of media, etc.) that would improve the water industry's communication with the media and with the greater public on trace organic compounds/ microconstituents? Similarly, what approaches have been effective? Why?

It would be helpful if talking points were coordinated between water and wastewater entities.

The industry could also be more proactive instead of only being reactive. For example, one water quality association recently prepared a story for the *SF Chronicle* about recycled water, talking about detections of endocrine disrupting compounds in almost all water supplies, the need to be protective of water and other key points. The article was very positive because the industry was in the driver's seat for article content. In general the wastewater industry tends to keep their heads low. This does not promote a positive image. If instead, the industry kept telling positive stories, they would balance out the negative press.

In addition, an approach that has really worked is doing regional positive media outreach. A local pollution prevention group has members from 43 wastewater agencies and we work together on regional projects such as regional thermometer exchanges, regional medicine disposal events, car wash outreach, etc. Utility 9 has a regional budget and can use it for advertising these events and trying to get local media coverage through a PR firm that consults with Utility 9. Again this is proactive, positive work with the media.

Interviews with Utility/Water District Teaming Partners
Utility 10

Date
July 28, 2008

Background
Rula Deeb provided a brief overview of the project, including the reduction of original scope, the focus on trace organic compounds and the four tasks within the scope of work. She explained that the team is now in the media analysis and review phase of the project (Task 3). After the call, Rula will send meeting notes summarizing Utility 10's answers back to them for review. The notes will then be finalized; utility names will be kept confidential in the final report. Utility 10 can also review the draft report and provide comments.

Questions and Responses
1. How important of a topic are trace organic compounds within your organization?

> They are pretty important, especially because Utility 10 practices recharge. Utility 10 has invested heavily in monitoring of recharge water for trace organic compounds. They received some calls from the public after the Associated Press (AP) article on pharmaceuticals. There were only 5 calls. Utility 10 residents did not reveal a strong concern over the issue of pharmaceuticals in drinking water.

2. Does your organization have a designated spokesperson to answer questions from the public about trace organic compounds? Also, does your organization have a designated spokesperson to answer questions from the media about trace organic compounds?

> Utility 10 has designated people to respond to the media, but they are not experts on trace organic compounds. The group is called CAPA (Communications and Public Affairs). CAPA's protocol is to get water quality and regulatory groups at Utility 10 involved to answer detailed questions. CAPA would take the responsibility of getting things out to the media.

3. Has your organization prepared any outreach materials on the issue of trace organic compounds? If so, please describe these materials and how they were distributed.

> No specific materials have been put together for the public on trace organic compounds. After the AP article, Utility 10 put together speaking points to be used by people within Utility 10 who are receiving calls from the public. Utility 10 used the same terminology to describe the compounds as AP (pharmaceuticals, endocrine-disrupting compounds). Utility 10 does not use one term to define the issue (e.g., trace organic compounds or microconstituents).
>
> Since the interest generated by the AP article, Utility 10 has initiated a drinking water monitoring program and prepared a press release describing the results. No pharmaceuticals were detected although caffeine was found in one of the treatment plant samples – the one that was not treated by granular activated carbon (GAC). In the press release, Utility 10 described the amount of water that a person would need to drink to get the equivalent caffeine that is in one cup of coffee. All testing was done in-house using analytical methods developed in conjunction with researchers and the State Department of Health laboratory. Utility 10 will send a copy of the press release to Linda, to be attached to the interview notes.

4. Have you ever used Water Environment Federation (WEF) or Water Environment Research Foundation's (WERF's) resources on trace organic compounds or microconstituents, such as their website, Technical Practice Updates or Question & Answers? If so, was the information helpful? What other resources have been useful?

> The water quality group has not used either the WEF or WERF resources on this topic. The water treatment group may have used these resources.

5. What, if any, are the principal concerns in your community related to trace organic compounds (e.g., human health concerns, environmental health concerns, other)? Please estimate the percentage of public concern associated with each.

> Based on the few calls that Utility 10 received, the public was concerned with human health. Some were also concerned about the health of their pets. Most asked for other information/references. Utility 10 provided information about what they knew about the issue: that there was incomplete health-related information, that these compounds had been detected at extremely low levels, and about advances in detection methods.

6. Has your organization developed any partnerships to assist in communicating the risks associated with trace organic compounds (e.g., partnerships with local research organizations, medical doctors, consultants)?

When the AP article came out, Utility 10 participated in a meeting with other adjacent cities, the Department of Health Services and representatives from the county to talk about ways to communicate with the public and discuss key messages. Everyone had their own thought processes about what should be done. A common theme that was discussed was the need for more communication with the public about health risks from the Health Department on this issue. It would be much better to get toxicologists and Public information officers from the Health Departments to come up with a press release about the health impacts of detecting these extremely low levels. It would be more effective to have this information come from a health department rather than from the local water and wastewater utilities. The county was going to take the lead on this but not much has happened to date.

Utility 10 has also worked collaboratively with researchers at Las Vegas regarding other aspects of trace organic compounds.

7. What tools, processes and approaches has your organization used to interact with the public? Describe the situations for which each approach has been used. Do you think that these methods attenuate public concern about trace organic compounds?

Utility 10 released a press statement about the results of their drinking water testing, in response to the AP article. There has been little public concern since that time. There is no specific information about this issue in their Consumer Confidence Report (CCR) or on their website. They will have to think about whether to report the drinking water test results in next year's CCR. If they received more public phone calls on this issue, Utility 10 would consider adding a website reference.

8. In your geographic area, have there been any changes in practices or programs to address trace organic compounds, such as monitoring programs, treatment studies or pharmaceutical take-back and disposal programs? Were any of these changes a direct result of public concern? Were any initiated by or supported by your organization?

After Utility 10 did testing, all the cities in the area got together and talked about the issue. Some of the other cities in the region chose to do testing also. Most of the conversations were about testing and testing protocols. Other cities have been more proactive and have started pharmaceutical take-back programs. Utility 10 has not started a take-back program – it would be through the environmental group but there is nothing about it on the environmental group's website.

9. Has your organization witnessed any changes in public involvement or perception regarding trace organics over time as a function of setting (e.g., during a drought, in response to an event, in response to the establishment of a program, etc.)? Please describe.

> There have not been any changes in treatment due to trace organics – GAC was added to deal with disinfection byproducts so any pharmaceuticals removal is just an added benefit. There has not been any evident increase in public concern about pharmaceuticals other than the response to the AP article.
>
> One caller complimented Utility 10 on the information provided in the CCR (not related to pharmaceuticals or personal care products). The caller particularly noted he had examined the CCR more carefully that year, in light of the recent media attention on bottled water and tap water. The caller found the information in the CCR helpful.
>
> Will people look for this issue in CCRs and expect information on whether pharmaceuticals and personal care products are detected? Time will tell.

10. Please identify a local media report (article, radio or TV report) as an example of how the media presents information on this issue and how your organization is represented.

> There may have been something in the local newspaper. Utility 10 can scan it to a pdf and email it to the project team.

11. Has your organization identified any communication deficiencies or needs (e.g., message content, style, trust, choice of media, etc.) that would improve the water industry's communication with the media and with the greater public on trace organic compounds/ microconstituents? Similarly, what approaches have been effective? Why?

> There needs to be more communication with the public to answer the question of what it all means, what health studies have been done, what studies are ongoing, what are the risks and dangers and to articulate what testing results mean.
>
> The meeting with other cities revealed a lack of experts to communicate with the media on health effects of these compounds, who has the necessary background, credibility and ability to help put the issue in context.

REFERENCES

Asmus, P., 2006. Test of a company's social pledges: Helping employees volunteer. Christian Science Monitor. October 20. Available online at http://www.csmonitor.com/2006/1019/p09s02-coop.html

ATSDR, 2008. ATSDR Glossary of Terms. Agency of Toxic Substances and Disease Registry. Available online at www.atsdr.cdc.gov/glossary.html

ATSDR, 2004. A Primer on Health Risk Communication Principles and Practices. Agency of Toxic Substances and Disease Registry. Available online at www.atsdr.cdc.gov/HEC/primer.html

AP, 2008. Study finds pharmaceutical drugs in tap water. Associated Press. March 10, 2008.

AwwaRF, 2003. Consumer Perceptions of Tap Water, Bottled Water, and Filtration Devices. American Water Works Research Foundation #2638.

AwwaRF, 2004. Risk Communication for Emerging Contaminants. American Water Works Research Foundation #2776.

AwwaRF, 2005. Message Management: Effective Communications. American Water Works Association Research Foundation #2766.

AwwaRF, 2008a. Guidance manual for communicating the value of water. American Water Works Association Research Foundation #3113.

AwwaRF, 2008b. State of knowledge of endocrine disruptors and pharmaceuticals in drinking water. American Water Works Association Research Foundation #3033.

Brown, L., 2004. Media relations for public safety professionals. Jones and Bartlett Publishers, Inc.

Cline, D., 1979. What are Effects? Living under Nuclear Fear. The Evening Sentinel. April 4.

Covello, V.T., 1991. Risk comparisons and risk communication: Issues and problems in comparing health and environmental risks. In R.E. Kasperson & P.J.M. Stallen (Eds.), Communicating risks to the public: International perspectives. The Netherlands: Kluwer Academic.

Covello, V.T., 2006. Core slides: Communicating in high concern/low trust situations: Applications to water security. Center for Risk Communication, NY.

Covello, V.T., 2007. Core slides: Risk communication. Presented at the Genesse County Health Department Public Health Week Conference: Healthy living, Healthy lifestyles, Flint, MI, April 4.

CRCWQT, date. In Theory and in Practice – Attitudes to Potential and Actual use of Recycled Water in and out of the House. (CRCWQT)

Drew, C.H., Grace, D.A., Silbernagel, S.M., Hemmings, E.S., Smith, A., Griffith, W.C., Takaro, T.K., and Faustman, E.M., 2003. Nuclear waste transportation: Case studies of identifying stakeholder risk information needs. Environmental Health Perspectives 111(3): 263-272.

Eriksen, K., 1990. Toxic reckoning: Business faces a new kind of fear. *Harvard Business Review, 68*, 118-126.

Fischhoff, B., Lichtenstein, S., Slovic, P., and Keeney, D., 1981. Acceptable risk. Cambridge, Massachusetts: Cambridge University Press.

Fischhoff, B., 1995. Risk perception and communication unplugged: Twenty years of process. Risk Analysis, 15, 137-145.

Flynn, J., Slovic, P., and Kunreuther, H. (Eds.), 2001. "Risk, Media and Stigma", Earthscan Publications Ltd, London, UK.

Flynn, J.H., Slovic, P., and Mertz, C.K. 1993. The Nevada Initiative: A risk communications fiasco. Risk analysis 13(5), 497-502.

Frazier-O'Brien, 2008. Contraceptives affect environment too, water expert tells conference. Catholic News Service/ USCCB. July 28.

Goldman, R. and Papson, S., 1998. Core cultural icons. Nike culture. Sage publications, London, UK.

Gregory, R., Flynn, J., and Slovic, P., 1995. Technological stigma. American Scientist 83, 220-223. Reprinted as Chapter 1 in Flynn, J. Slovic, P., and Kunreuther, H. (Eds.), 2001. "Risk, Media and Stigma", Earthscan Publications Ltd., London, UK.

Harris Poll, 2005. Three-quarters of U.S. adults agree environmental standards cannot be too high and continuing improvements must be made regardless of cost. Available online at www.harrisinteractive.com/harris_poll/index.asp?PID=607

Health Canada, 2006. A Framework for Strategic Risk Communications within the Context of Health Canada and the PHAC's Integrated Risk Management. Chapter 2 in Strategic Risk Communications Framework for Health Canada and the Public Health Agency of Canada. ISBN No. 0-662-44596-1. November. Available online at www.riskcommunications.gc.ca

Jeffrey, P. and Jefferson, B., 2003. Public receptivity regarding in-house water recycling: Results from a UK survey. Water Science and Technology: Water Supply. International Water Association World Water Congress No. 3, Melbourne, Australia 3(3): 109-116. ISBN 1-84339-445-6.

Joseph, P., 2008. Numberstruck. California. May/June 2008. Available online at http://alumni.berkeley.edu/california/200805/praxisnumber.asp

Kahney, L., 2002. Apple: It's all about the brand. WIRED commentary, December 4. Available online at http://www.wired.com/gadgets/mac/commentary/cultofmac/2002/12/56677

Kasperson, J.X., Kasperson, R.E., Pidgeon, N., and Slovic, P., 2003. The social amplification of risk: Assessing fifteen years of research and theory. Chapter 1 in The social amplification of risk, Eds. Pidgeon, N., Kasperson, R.E and Slovic, P. Cambridge University Press, ISBN 0521520444, 9780521520447, 448 pages.

Kavanaugh, M., 2008. Personal communication with Michael C. Kavanaugh regarding the Stringfellow site, Glen Avon, CA.

Keller, K.L., 2000. The brand report card. Harvard Business Review, 78(1): 3-10.

Kraus, N., Malmfors, T., and Slovic, P., 1992. Intuitive Toxicology: Expert and Lay Judgments of Chemical Risks. Risk Analysis, 12: 215-232.

Kunreuther, H. and Slovic, P., 2000. Coping with stigma: Challenges and opportunities. In Flynn, J. Slovic, P. and Kunreuther, H. (Eds.), 2001. *Risk, media and stigma* (pp. 331-352). London: Earthscan, p. 341.

Leiss, W.. 2001. Dioxins or chemical stigmata. Chapter 16 in Flynn, J. Slovic, P., and Kunreuther, H. (Eds.), 2001. "Risk, Media and Stigma", Earthscan Publications Ltd, London, UK.

Long, R., 2001. An industry perspective on risk and stigma. Chapter 17 in Flynn, J. Slovic, P.. and Kunreuther, H. (Eds.), 2001. "Risk, Media and Stigma", Earthscan Publications Ltd, London, UK.

Morgan, M.G., Fischhoff, B., Bostrom, A., Lave, L., and Atman, C.J., 2002. Risk communication: A mental models approach. New York: Cambridge University Press.

NACWA, 2005. Pharmaceuticals and Personal Care Products in the Environment: A White Paper on Options for the Wastewater Treatment Community. National Association of Clean Water Agencies. November.

NORC, 2004. General Social Survey. National Opinion Research Center, Chicago, IL. Available online at http://www.norc.org/GSS+Website/.

NRC, 1996. Understanding risk: Informing decisions in a democratic society. Stern, P.C. and Fineburg, H.V. (Eds.), Committee on risk characterization, National Research Council, ISBN 0-309-57849-3.

OEHHA, 2006. Public health goal for N-nitrosodimethylamine (NDMA) in drinking water. Prepared by the Pesticide and Environmental Toxicology Branch, Office of Environmental Health and Hazard Assessment, California Environmental Protection Agency. December.

Perras, J., 2005. Communications Essentials for Agency Attorneys and Managers. Perras and Associates. NACWA Developments in Clean Water Law Seminar, November 10, Santa Fe, NM.

Purchase, I.F.H. and Slovic, P., 1999. Quantitative risk assessment breeds fear. Human and Ecological Risk Assessment, 5(3): 445-453.

Recycled Water Task Force, 2003. Water recycling, 2030. Recommendations of California's Recycled Water Task Force.

Rogers, Everett M., 1995. Diffusion of Innovations. Simon & Schuster.

Roger, P., Bhatia, R., and Huber, A., 1998. Water as a social and economic good: How to put the principle into practice. Stockholm, Sweden: Global Water Partnership/Swedish International.

Roseth, N., 2006. Community Views on Water Shortages and Conservation. Research Report No. 28. The Cooperative Research Centre (CRC) for Water Quality and Treatment, Australia.

Roseth, N., 2008. Community Views on Recycled Water – The Impact of Information. Research Report No. 48. The Cooperative Research Centre (CRC) for Water Quality and Treatment, Australia.

Rozin, P., 2001. Technological stigma: Some perspectives from the study of contagion. Chapter 3 in Flynn, J. Slovic, P. and Kunreuther, H. (Eds.), 2001. "Risk, Media and Stigma", Earthscan Publications Ltd, London, UK.

Rozin, P., Haidt, J., and McCauley, C.R., 2000. Disgust. In Handbook of emotions, Eds. M. Lewis and J.M.Haviland-Jones. Second edition, Guilford Press, New York, pp. 637-653.

Rozin, P., Ashmore, M., and Markwith, M., 1996. Lay American conceptions of nutrition: Dose insensitivity, categorical thinking, contagion and the monotonic mind. Health Psychology, 15(6): 438-447. November.

Rosen, J.D., 1990. Much ado about Alar. Issues in Science and Technology 7(1): 85-90

Sandman, P.M., 2004. Three Mile Island – Twenty-five years later. Safety at Work. April 24. Available online at www.psandman.com/articles/3-mile2.pdf.

Sandman, P., 2006. Question 5 of 2006 Survey on Globalization, Corporate Responsibility & Sustainability. Available online at http://www.sustainability.com/network/2006_responses_home.asp

Satterfield, T., Slovic, P., Gregory, R., Flynn, J., and Mertz, C.K., 2001. Risk lived, stigma experienced: Reflections on the limits of adaptations. Chapter 5 in Flynn, J. Slovic, P. and Kunreuther, H. (Eds.), 2001. "Risk, Media and Stigma", Earthscan Publications Ltd, London, UK.

Slovic, P., 1993. Perceived risk, trust, and democracy. Risk Analysis 13(6): 675-682.

Slovic, P., 1999. Trust, emotion, sex, politics and science: Surveying the risk-assessment battlefield. Risk Analysis, 19(4): 689-701.

Slovic, P., Finucane, M. L., Peters, E., and MacGregor, D.G., 2004. Risk as analysis and risk as feelings: Some thoughts about affect, reason, risk, and rationality. Risk Analysis, 24(2), 1-12.

Slovic, P., 2007. Risk Perception and Communication (Lessons from the Risk Assessment Battlefield). Water Environment Research Foundation Workshop on Trace Organic Compounds: Mapping a Collaborative Research Roadmap, San Francisco, CA, May 17.

Slovic, P., Peters, E., Grana, J., Berger, S., and Dieck, G.S., 2007. Risk perception of prescription drugs: Results of a national survey. Drug Information Journal 41: 81-100.

TIME magazine, 1979. A nuclear nightmare. Monday, April 9. Available online at www.time.com/time/printout/0,8816,920196,00.html

Van Ryan, A.D., 2003. Managing trust by bold inclusion. Environmental Law Newsletter, Fall.

Walker, 2001. Defining and identify "stigma". Chapter 22 in Flynn, J. Slovic, P. and Kunreuther, H. (Eds.), 2001. "Risk, Media and Stigma", Earthscan Publications Ltd, London, UK.

WateReuse Association, 2004. Innovative Applications in Water Reuse: Ten Case Studies.

WateReuse Foundation, 2004a. Best Practices for Developing Indirect Potable Reuse Projects: Phase 1 Report. Product No. 01-004-01.

WateReuse Foundation, 2004b. Research Needs Assessment Workshop: Human Reactions to Water Reuse. Prepared by Brent M. Haddad, University of California at Santa Cruz. Co-sponsored by the U.S. Bureau of Reclamation. WateReuse Foundation Report WRF-03-011

WateReuse Foundation, 2006. Marketing Nonpotable Recycled Water: A Guidebook for Successful Public Outreach & Customer Marketing. Product No. 03-005-01.

WateReuse Foundation, 2008. Request for proposals. Talking about water: Vocabulary and images that support informed decisions about water recycling and desalination. WRF-07-03. Issued May 9.

WEF, 2007a. Microconstituents in the water environment. Resource paper for public and press. WEF's Microconstituents Community of Practice of the Technical Practice Committee.

WEF, 2007b. Microconstituents glossary. Prepared by Microconstituents Community of Practice of the Technical Practice Committee.

WERF, 2003. Water Reuse: Understanding Public Perception and Participation. Water Environment Research Foundation, Final Report 00PUM1.

WERF, 2004. Public Perception of Biosolids Recycling: Developing Public Participation and Earning Trust. Water Environment Research Foundation, Final Report 00PUM5.

WERF, 2007. Proceedings from the WERF Workshop on Trace Organic Compounds: Mapping a Collaborative Research Roadmap. May 17-18, San Francisco, CA. Available online at http://www.werf.org/Content/NavigationMenu/Archive/OurResearchProcess/CECWorkshop/ResearchRoadmap/default.htm

WERF, 2008. Technical brief: Trace organic compounds and implications for wastewater treatment. Water Environment Research Foundation, Report CEC3R07.

WERF, 2008b. Q&A on pharmaceuticals in water. Available online at www.werf.org/AM/Template.cfm?Section=Home&TEMPLATE=/CM/ContentDisplay.cfm&CONTENTID=7239

WHO, 2007. Effective media communication during public emergencies: A WHO handbook. Geneva, Switzerland. July. ISBN-13: 9789241547031; ISBN-10: 9241547030.

WordNet, 2008. "Trust" keyword search. Available online at http://wordnet.princeton.edu/

WASTEWATER UTILITY

Alabama
Montgomery Water Works & Sanitary Sewer Board

Alaska
Anchorage Water & Wastewater Utility

Arizona
Avondale, City of
Glendale, City of, Utilities Department
Mesa, City of
Peoria, City of
Phoenix Water Services Dept.
Pima County Wastewater Management
Safford, City of
Tempe, City of

Arkansas
Little Rock Wastewater Utility

California
Central Contra Costa Sanitary District
Corona, City of
Crestline Sanitation District
Delta Diablo Sanitation District
Dublin San Ramon Services District
East Bay Dischargers Authority
East Bay Municipal Utility District
El Dorado Irrigation District
Fairfield-Suisun Sewer District
Fresno Department of Public Utilities
Inland Empire Utilities Agency
Irvine Ranch Water District
Las Gallinas Valley Sanitary District
Las Virgenes Municipal Water District
Livermore, City of
Los Angeles, City of
Los Angeles County, Sanitation Districts of
Napa Sanitation District
Novato Sanitary District
Orange County Sanitation District
Palo Alto, City of
Riverside, City of
Sacramento Regional County Sanitation District
San Diego Metropolitan Wastewater Department, City of
San Francisco, City & County of
San Jose, City of
Santa Barbara, City of
Santa Cruz, City of
Santa Rosa, City of
South Bayside System Authority
South Coast Water District
South Orange County Wastewater Authority
Stege Sanitary District
Sunnyvale, City of
Union Sanitary District
West Valley Sanitation District

Colorado
Aurora, City of
Boulder, City of
Greeley, City of
Littleton/Englewood Water Pollution Control Plant
Metro Wastewater Reclamation District, Denver

Connecticut
Greater New Haven WPCA
Stamford, City of

District of Columbia
District of Columbia Water & Sewer Authority

Florida
Broward, County of
Fort Lauderdale, City of
Jacksonville Electric Authority (JEA)
Miami-Dade Water & Sewer Authority
Orange County Utilities Department
Pinellas, County of
Reedy Creek Improvement District
Seminole County Environmental Services
St. Petersburg, City of
Tallahassee, City of
Tampa, City of
Toho Water Authority
West Palm Beach, City of

Georgia
Atlanta Department of Watershed Management
Augusta, City of
Clayton County Water Authority
Cobb County Water System
Columbus Water Works
Fulton County
Gwinnett County Department of Public Utilities
Savannah, City of

Hawaii
Honolulu, City & County of

Idaho
Boise, City of

Illinois
Decatur, Sanitary District of
Greater Peoria Sanitary District
Kankakee River Metropolitan Agency
Metropolitan Water Reclamation District of Greater Chicago
Wheaton Sanitary District

Iowa
Ames, City of
Cedar Rapids Wastewater Facility
Des Moines, City of
Iowa City

Kansas
Johnson County Wastewater
Lenexa, City of
Unified Government of Wyandotte County/ Kansas City, City of

Kentucky
Louisville & Jefferson County Metropolitan Sewer District

Louisiana
Sewerage & Water Board of New Orleans

Maine
Bangor, City of
Portland Water District

Maryland
Anne Arundel County Bureau of Utility Operations
Howard County Bureau of Utilities
Washington Suburban Sanitary Commission

Massachusetts
Boston Water & Sewer Commission
Massachusetts Water Resources Authority (MWRA)
Upper Blackstone Water Pollution Abatement District

Michigan
Ann Arbor, City of
Detroit, City of
Holland Board of Public Works
Saginaw, City of
Wayne County Department of Environment
Wyoming, City of

Minnesota
Rochester, City of
Western Lake Superior Sanitary District

Missouri
Independence, City of
Kansas City Missouri Water Services Department
Little Blue Valley Sewer District
Metropolitan St. Louis Sewer District

Nebraska
Lincoln Public Works and Utilities Department

Nevada
Henderson, City of
Las Vegas, City of
Reno, City of

New Jersey
Bergen County Utilities Authority
Ocean County Utilities Authority
Passaic Valley Sewerage Commissioners

New York
New York City Department of Environmental Protection

North Carolina
Charlotte/Mecklenburg Utilities
Durham, City of
Metropolitan Sewerage District of Buncombe County
Orange Water & Sewer Authority
University of North Carolina, Chapel Hill

Ohio
Akron, City of
Butler County Department of Environmental Services
Columbus, City of
Metropolitan Sewer District of Greater Cincinnati
Montgomery, County of
Northeast Ohio Regional Sewer District
Summit, County of

Oklahoma
Oklahoma City Water & Wastewater Utility Department
Tulsa, City of

Oregon
Albany, City of
Clean Water Services
Eugene, City of
Gresham, City of
Portland, City of Bureau of Environmental Services
Water Environment Services

Pennsylvania
Hemlock Municipal Sewer Cooperative (HMSC)
Philadelphia, City of
University Area Joint Authority

South Carolina
Charleston Water System
Mount Pleasant Waterworks & Sewer Commission
Spartanburg Water

Tennessee
Cleveland Utilities
Murfreesboro Water & Sewer Department
Nashville Metro Water Services

Texas
Austin, City of
Dallas Water Utilities
Denton, City of
El Paso Water Utilities

Fort Worth, City of
Houston, City of
San Antonio Water System
Trinity River Authority

Utah
Salt Lake City Corporation

Virginia
Alexandria Sanitation Authority
Arlington, County of
Fairfax County
Hampton Roads Sanitation District
Hanover, County of
Henrico, County of
Hopewell Regional Wastewater Treatment Facility
Loudoun Water
Prince William County Service Authority
Richmond, City of
Rivanna Water & Sewer Authority

Washington
Everett, City of
King County Department of Natural Resources
Seattle Public Utilities
Sunnyside, Port of
Yakima, City of

Wisconsin
Green Bay Metro Sewerage District
Kenosha Water Utility
Madison Metropolitan Sewerage District
Milwaukee Metropolitan Sewerage District
Racine, City of
Sheboygan Regional Wastewater Treatment
Wausau Water Works

Australia
ACTEW (Ecowise)
South Australian Water Corporation
South East Water Limited
Sydney Water Corporation
Water Corporation of Western Australia

Canada
Edmonton, City of/Edmonton Waste Management Centre of Excellence
Lethbridge, City of
Regina, City of, Saskatchewan
Toronto, City of, Ontario
Winnipeg, City of, Manitoba

New Zealand
Watercare Services Limited

United Kingdom
Anglian Water Services, Ltd.

STORMWATER UTILITY

California
Fresno Metropolitan Flood Control District
Los Angeles, City of, Department of Public Works
Monterey, City of
San Francisco, City & County of
Santa Rosa, City of
Sunnyvale, City of

Colorado
Aurora, City of
Boulder, City of

Florida
Orlando, City of

Iowa
Cedar Rapids Wastewater Facility
Des Moines, City of

Kansas
Overland Park, City of

Kentucky
Louisville & Jefferson County Metropolitan Sewer District

Maine
Portland Water District

North Carolina
Charlotte, City of, Stormwater Services

Pennsylvania
Philadelphia, City of

Tennessee
Chattanooga Stormwater Management

Texas
Harris County Flood Control District, Texas

Washington
Bellevue Utilities Department
Seattle Public Utilities

STATE

Connecticut Department of Environmental Protection
Kansas Department of Health & Environment
New England Interstate Water Pollution Control Commission (NEIWPCC)
Ohio River Valley Sanitation Commission
Urban Drainage & Flood Control District, CO

CORPORATE

ADS Environmental Services
Advanced Data Mining International
AECOM
Alan Plummer & Associates
Alpine Technology Inc.
Aqua-Aerobic Systems Inc.
Aquateam–Norwegian Water Technology Centre A/S
ARCADIS
Associated Engineering
Bernardin Lochmueller & Associates
Black & Veatch
Blue Water Technologies, Inc.
Brown & Caldwell
Burgess & Niple, Ltd.
Burns & McDonnell
CABE Associates Inc.
The Cadmus Group
Camp Dresser & McKee Inc.
Carollo Engineers Inc.
Carpenter Environmental Associates Inc.
CET Engineering Services
CH2M HILL
CRA Infrastructure & Engineering
CONTECH Stormwater Solutions
D&B/Guarino Engineers, LLC
Damon S. Williams Associates, LLC
Ecovation
EMA Inc.
Environmental Operating Solutions, Inc.
Environ/The ADVENT Group, Inc.
Fay, Spofford, & Thorndike Inc.
Freese & Nichols, Inc.
ftn Associates Inc.
Gannett Fleming Inc.
Garden & Associates, Ltd.
Geosyntec Consultants
GHD
Greeley and Hansen LLC
Hazen & Sawyer, P.C.
HDR Engineering Inc.
HNTB Corporation
Hydromantis Inc.
HydroQual Inc.
Infilco Degremont Inc.
Jacques Whitford NAWE, Inc.
Jason Consultants LLC Inc.
Jordan, Jones, & Goulding Inc.
KCI Technologies Inc.
Kelly & Weaver, P.C.
Kennedy/Jenks Consultants
KMK Consultants
Larry Walker Associates
Limno-Tech Inc.
The Low Impact Development Center Inc.
Malcolm Pirnie Inc.
Material Matters
McKim & Creed
MPR Engineering Corporation, Inc.
MWH
NTL Alaska, Inc.
O'Brien & Gere Engineers Inc.
Odor & Corrosion Technology Consultants Inc.
Paradigm Environmental Technologies, Inc.
Parametrix Inc.
Parsons
Post, Buckley, Schuh & Jernigan
Praxair, Inc.
Ring Industrial Group
RMC Water & Environment
Ross & Associates Ltd.
SAIC
Siemens Water Technologies
The Soap & Detergent Association
Smith & Loveless, Inc.
Southeast Environmental Engineering, LLC
Stearns & Wheler, LLC
Stone Environmental Inc.
Stratus Consulting Inc.
Synagro Technologies Inc.
Tetra Tech Inc.
Trojan Technologies Inc.
Trussell Technologies, Inc.
Uni-Bell PVC Pipe Association
URS Corporation
Westin Engineering Inc.
Wright Water Engineers
Zoeller Pump Company

INDUSTRY

American Electric Power
American Water
Anglian Water Services, Ltd.
Chevron Energy Technology
The Coca-Cola Company
Dow Chemical Company
DuPont Company
Eastman Chemical Company
Eli Lilly & Company
Johnson & Johnson
Merck & Company Inc.
Procter & Gamble Company
Suez Environnment
United Utilities North West (UUNW)
United Water Services LLC

Note: List as of 2/21/09

WERF Board of Directors

WERF Research Council

IWA Publishing's authorised EU representative for General Product Safety Regulations is Diane D'Arras, 15 rue Duret, 75116 Paris, France, e-mail: safety@iwap.co.uk.

Printed and bound by CPI Group (UK) Ltd, Croydon, CR0 4YY
12/05/2026
02108892-0015